Checklist

Issues of Web Design

Issue	Characteristics	Guidelines
Writing	Text	• Chunk information for easy access and quick reading. (72)
		• Consider audience needs. (72)
		• Create chunks (modules) for multiple purposes. (72)
	Nonlinear readers (hypertext)	• Write each chunk to be understandable regardless of order or context. (72)
	Broad audience	• Define all terms and provide necessary links. (73)
		• Provide a feedback and question mechanism. (73)
		• Write short sentences to facilitate reading and translation. (73)
Design	Organization	• Use top-down organization. (73)
		• Structure links by hierarchy of importance. (73)
	Typography	• Follow principles of good type design: Use clear contrast between fonts; avoid using all capital letters; use a serif font for body text and sans serif for display text. (74)
		• Use fonts designed for computer screen reading (Georgia, Verdana). (74)
	Line length	• Chunk information into small units. (75)
		• Use short text lines. (75)
		• Avoid right-justified text. (75)
	Appearance	• Focus on cultural and organizational considerations. (171)
		• Avoid mixing too many colors, and use colors appropriate to the audience. (75)
Technical features	Accessibility	• Consider the technical capabilities of audience's computers. (75)
		• Use standard file formats. (77)
		• Design for industry standard browsers. (77)
		• Ensure that all links are correct. (77)
		• Avoid gratuitous use of bells and whistles. (77)
	Usability	• Test site to ensure it is functional and usable. (77)

(Numbers in parentheses refer to the first page of major discussion in the text.)

A Concise Guide to
TECHNICAL
COMMUNICATION

A Concise Guide to

Technical

Communication

Laura J. Gurak
University of Minnesota

John M. Lannon
University of Massachusetts–Dartmouth

Longman

New York   San Francisco   Boston
London   Toronto   Sydney   Tokyo   Singapore   Madrid
Mexico City   Munich   Paris   Cape Town   Hong Kong   Montreal

Editor-in-Chief: Joseph Terry
Acquisitions Editor: Lynn M. Huddon
Development Editor: David Munger
Marketing Manager: Carlise Paulson
Supplements Editor: Donna Campion
Production Manager: Bob Ginsberg
Project Coordination, Text Design, and Electronic Page Makeup: Nesbitt Graphics, Inc.
Cover Design Manager and Cover Designer: John Callahan
Senior Manufacturing Buyer: Dennis Para
Printer and Binder: Von Hoffmann Press, Inc.
Cover Printer: Von Hoffmann Press, Inc.

For permission to use copyrighted material, grateful acknowledgment is made to the copyright holders on pp. 329–330, which are hereby made part of this copyright page.

Between the time Web site information is gathered and published, some sites may have closed. Also, the transcription of URLs can result in typographical errors. The publisher would appreciate notification where these occur so that they may be corrected in subsequent editions.

Many of the designations used by manufacturers and sellers to distinguish their products are claimed as trademarks. Where these designations appear in this book, and Addison Wesley Longman was aware of a trademark claim, the designations have been printed in initial caps.

Library of Congress Cataloging-in-Publication Data

Gurak, Laura. J.
    A concise guide to technical communication/Laura J. Gurak, John M. Lannon—1st. ed.
        p. cm.
    Includes bibliographical references and index.
    ISBN 0-321-06106-3
    I. Communication of technical information. I. Lannon, John M. II. Title.

    T10.5.G83 2001
    601'.4--dc21

                                                            00-046176

Please visit our website at http://www.ablongman.com/gurak

ISBN 0-321-06106-3

1 2 3 4 5 6 7 8 9 10—VH—03 02 01 00

Brief Contents

PART 1   **TECHNICAL COMMUNICATION TECHNIQUES AND CONSIDERATIONS   1**

CHAPTER 1   *An Introduction to Technical Communication   3*

CHAPTER 2   *A World of People and Purposes   14*

CHAPTER 3   *Designing Usable Information   30*

CHAPTER 4   *Performing Research for Technical Communication   50*

CHAPTER 5   *Technical Communication in a Digital World   70*

CHAPTER 6   *Ethical Issues in Technical Communication   86*

CHAPTER 7   *Copyright and Privacy   104*

CHAPTER 8   *Page Layout and Document Design   120*

CHAPTER 9   *Graphics and Visual Information   147*

PART 2   **TECHNICAL COMMUNICATION SITUATIONS AND APPLICATIONS   181**

CHAPTER 10   *Everyday Communication Situations   183*

CHAPTER 11   *Product-Oriented Communication Situations   215*

CHAPTER 12   *Complex Communication Situations   246*

Appendix A   *Grammar   285*

Appendix B   *Documenting Sources   302*

Detailed Contents

Preface   xxi

PART 1   TECHNICAL COMMUNICATION TECHNIQUES AND CONSIDERATIONS   1

CHAPTER 1
An Introduction to Technical Communication   3

Communicating about Technology   4
Main Characteristics of Technical Communication   4
Accessibility   4
Usability   5
Relevance   5
Types of Technical Communication   7
Technical Communication in the Workplace   9
Technical Communication in an Electronic Age   9
Societal Dimensions of Technical Communication   10
Ethical Dimensions of Technical Communication   11
Review Checklist   11
Exercises   12
The Collaboration Window   12
The Global Window   13
Click on This   13

CHAPTER 2
A World of People and Purposes   14

People and Purposes   15
Analyzing Your Audience   18
Analyzing the Communication Purpose   19

Analyzing the Communication Context   22

Conducting an Audience/Purpose Interview   23

More Tools for Understanding Audience   24

Using Information from Your Analysis   25

Audiences Are Not Passive   25

Typical Audiences and Purposes for Technical
   Communication   26
Review Checklist   27
Exercises   28
The Collaboration Window   28
The Global Window   29
Click on This   29

chapter 3
Designing Usable Information   30

Usability and Technical Information   31

Usability—During the Planning Stages   31
Perform an Audience and Purpose Analysis   31
Perform a Task Analysis   31
Develop an Information Plan   33
Do the Research   33

Usability—During the Writing and Design Process   33
Test Early Versions of Your Communication Project   33
Revise Your Plan and Your Product   35
Create Layered Communication   35

Usability—After the Information Is Released   35
Provide Mechanisms for User Feedback   37
Plan for the Next Version or Release   37

Writing and Organizing Information for Usability   37
Using Appropriate Grammar and Style   37
Creating an Overview   43
Chunking Information   43
Creating Headings   45
Using the Margins for Commentary   45
Review Checklist   46
Exercises   46
The Collaboration Window   48

The Global Window   49
Click on This   49

chapter 4
Performing Research for Technical Communication   50

Thinking Critically About Research   51
Primary Research   51
  Informative Interviews   51
  Surveys and Questionnaires   52
  Public Records and Organizational Publications   53
  Personal Observation and Experiments   54
Internet Research   54
  Usenet News   54
  Listservs   55
  Electronic Magazines (Zines)   56
  Email Inquiries   56
  The Web   57
Other Electronic Research Tools   60
  Compact Discs   60
  Online Retrieval Services   61
  Card Catalog   62
Hard Copy Research   62
  Bibliographies   62
  Encyclopedias   62
  Dictionaries   63
  Handbooks   63
  Almanacs   63
  Directories   63
  Guides to Literature   63
  Indexes   63
  Abstracts   65
  Access Tools for U.S. Government Publications   65
  Microforms   65
Review Checklist   66
Exercises   66
The Collaboration Window   67
The Global Window   68
Click on This   68

chapter 5
Technical Communication in a Digital World   70

Communicating in Digital Space   71
Technical Commmunication Principles Still Apply   71
Designing Information for the New Media   71
    Writing Issues   72
    Design Issues   73
    Technical Issues   75

Online Documentation and Interface Design   77
    Online Documentation   77
    Interface Design   77

Corresponding Over the Wires: Email Conventions   79
Working on the Wires   81
Presentation Software   81
Review Checklist   82
Exercises   83
The Collaboration Window   84
The Global Window   84
Click on This   85

chapter 6
Ethical Issues in Technical Communication   86

Ethics, Technology, and Communication   87
Case Examples of Ethical Issues in Technical
    Communication   87
    Three Mile Island   87
    The Space Shuttle Challenger   88
    The Pentium III Chip   89

Types of Ethical Choices   90
    Kant's Categorical Imperative   90
    Utilitarianism   91
    Ethical Relativism   91

What Is Legal Is Not Always Ethical   92
    Copyright   92
    Plagiarism   93
    Privacy   93

Additional Ways in Which Actions Can Be Unethical   94

    Yielding to Social Pressure   95

    Mistaking Groupthink for Teamwork   95

    Suppressing Knowledge the Public Needs   95

    Exaggerating Claims about Technology   96

    Exploiting Cultural Differences   96

Types of Technical Communication Affected by Ethical Issues   96

    Graphics   96

    Web Pages and the Internet   97

    Memos   97

    Instructions   97

    Reports   98

    Proposals   98

    Oral Presentations   98

Responding to Ethical Situations   99

Review Checklist   101

Exercises   102

The Collaboration Window   102

The Global Window   102

Click on This   103

chapter 7
Copyright and Privacy   104

Why Technical Communicators Need to Understand Copyright and Privacy   105

Section One: Copyright

Copyright—An Overview   105

    Copyright Differs from Patent or Trademark Law   106

    How Individuals and Companies Establish Their Copyright   107

    What Rights a Copyright Holder Can Claim   107

    When You Can and Cannot Use Copyrighted Material   107

    When Your Company Owns the Material   110

Documenting Your Sources   110

Electronic Technologies and Copyright   111

    Photocopiers and Scanners   111

The Web as a Marketplace of Ideas and Information   111
Using Material from the Internet   112
Locating Copyright-Free Clip Art   112
Email and Electronic Messages   113
CD-ROMs and Multimedia   113

Section Two: Privacy

Privacy—An Overview   113

Computer Technologies, Privacy, and Technical
    Communication   114
Privacy in Cyberspace   114
Privacy and Documentation   116
Privacy and Videotapes   117
Review Checklist   117
Exercises   118
The Collaboration Window   118
The Global Window   119
Click on This   119

chapter 8
Page Layout and Document Design   120

Creating Visually Effective Documents   121

Typography   121
Fonts and Families   122
A Typeface Sends a Message   125
Certain Fonts Go Together   125
Choosing a Type Size for Readability   127
Using Customary Formats for Specific Purposes   127

Page Layout   128
How Readers View a Page   128
Using a Grid Structure   129
Creating Areas of Emphasis   129
Using White Space   130
Using Lists   130
Using Headings   131

Creating an Effective Table of Contents and Index   137

Creating and Using Style Sheets   137

Organizational Style Guides   137

Using Word-Processing and Page Layout Software   139
  Word-Processing and Page Layout Software   139
  Markup Languages   139

Designing Electronic Documents   142
  Web Pages   142
  Online Help   143
  CD-ROMS   143

Review Checklist   143
Exercises   144
The Collaboration Window   145
The Global Window   145
Click on This   146

chapter 9
Graphics and Visual Information   147

The Power of the Picture   148

When to Use Visuals   149

Different Visuals for Different Audiences   149

Text into Tables   150

Numbers into Images   151
  Graphs   154
  Charts   157

Illustrations   161

Diagrams   162
  Exploded Diagrams   163
  Cutaway Diagrams   163

Symbols and Icons   165

Wordless Instruction   166

Photographs   166

Maps   166

Visualization and Medical Imaging   169

Software and Web-Based Images   169

Using Color   170

Avoiding Visual Noise   172

Visuals and Ethics   172
Cultural Considerations   172
Review Checklist   173
Exercises   178
The Collaboration Window   178
The Global Window   179
Click on This   179

PART 2   Technical Communication Situations and Applications   181

chapter 10
Everyday Communication Situations   183

Email   184
The Situation   184
Audience and Purpose Analysis   184
Types of Email   184
Typical Components of Email   185
Usability Considerations   186

Memos   187
The Situation   187
Audience and Purpose Analysis   187
Types of Memos   187
Typical Components of a Memo   189
Usability Considerations   189

Letters   190
The Situation   190
Audience and Purpose Analysis   190
Types of Letters   191
Typical Components of a Letter   192
Specialized Components of a Letter   197
Usability Considerations   198

Short Reports   199
The Situation   199
Audience and Purpose Analysis   200
Types of Short Reports   200
Typical Components of Short Reports   201
Usability Considerations   201

ORAl COMMUNICATION   206
 The Situation   206
 Audience and Purpose Analysis   206
 Types of Oral Presentations   207
 Typical Components of Oral Presentations   208
 Usability Considerations   210
Review Checklist   212
Exercises   212
The Collaboration Window   213
The Global Window   214
Click on This   214

chapter 11
Product-Oriented Communication Situations   215

Specifications   216
 The Situation   216
 Audience and Purpose Analysis   216
 Types of Specifications   218
 Typical Components of Specifications   218
 Usability Considerations   221

Brief Instructions   222
 The Situation   222
 Audience and Purpose Analysis   222
 Types of Brief Instructions   222
 Typical Components of Instructions   225
 Usability Considerations   225

Procedures   226
 The Situation   226
 Audience and Purpose Analysis   227
 Types of Procedures   227
 Typical Components of Procedures   228
 Usability Considerations   231

Documentation and Manuals   232
 The Situation   232
 Audience and Purpose Analysis   232
 Types of Manuals   233
 Typical Components of Manuals   233
 Usability Considerations   235

TECHNICAL MARKETING MATERIAL   236

The Situation   236

Audience and Purpose Analysis   238

Types of Technical Marketing Material   238

Typical Components of Technical Marketing Material   240

Usability Considerations   240

REVIEW CHECKLIST   243

EXERCISES   244

THE Collaboration WINDOW   244

THE Global WINDOW   244

Click ON This   245

CHAPTER 12

Complex Communication Situations   246

DEFINITIONS AND DESCRIPTIONS   247

The Situation   247

Audience and Purpose Analysis   247

Types of Definitions and Descriptions   255

Typical Components of Definitions and Descriptions   255

Usability Considerations   256

LONG REPORTS   258

The Situation   258

Audience and Purpose Analysis   259

Types of Long Reports   259

Typical Components of Long Reports   269

Usability Considerations   271

PROPOSALS   272

The Situation   272

Audience and Purpose Analysis   272

Types of Proposals   273

Typical Components of Proposals   277

Usability Considerations   278

REVIEW CHECKLIST   280

EXERCISES   280

THE Collaboration WINDOW   284

The Global Window   284
Click on This   284

Appendix A
Grammar   285

GRAMMAR ISSUES   285
Punctuation   286
Lists   288
Avoiding Sentence Fragments and Run-On
    Sentences   290
Usage (Commonly Misused Words)   293
Subject-Verb and Pronoun-Antecedent Agreement   294
Faulty Modification   295
Mechanics   298

Appendix B
Documenting Sources   302

QUOTING THE WORK OF OTHERS   303
Paraphrasing the Work of Others   303
What You Should Document   303

HOW YOU SHOULD DOCUMENT   304

MLA DOCUMENTATION STYLE   305
MLA Parenthetical References   305
MLA Works Cited Entries   306
MLA Works Cited Page   316

APA DOCUMENTATION STYLE   316
APA Parenthetical References   316
APA Reference List Entries   317
APA Sample References List   324

References   325
Credits   329
Index   331

Preface

Even the most casual observer can see the powerful, compelling relationships between technology and communication. The growing number and complexity of new technologies, from personal computers to medical devices to Internet applications, require users to have accurate information on how to use and maintain these devices. The use of technologies—HTML coding, Web applications, online help screens—to communicate information across global boundaries means that every professional must understand technology.

Ever increasing is the need for teaching technical communication to future professional communicators as well as future technical experts who must communicate their knowledge to others. To meet this need, institutions offer technical communication degrees or certificates through a combination of face-to-face, interactive television, or Web-based classes. Some students major or minor in technical communication while others take technical writing and communication courses to fulfill humanities requirements and enhance their skills as engineers, scientists, or other specialists. In addition, writing-intensive programs, especially in engineering or science institutions, often focus on technical writing. Even high schools are now adding technical writing to their list of elective classes.

Whatever its context, technical communication is rarely a value-neutral exercise in "information transfer." It is a rhetorical, social transaction comprising interpersonal, cultural, ethical, legal, and technological components. In today's global environment, a one-size-fits-all approach simply does not work. Effective technical communication must be clear, accurate, and organized, and must be tailored for specific audiences and purposes.

With those requirements in mind we have created this first edition of the *Concise Guide to Technical Communication.* This book draws on the strengths of John M. Lannon's best-selling *Technical Communication* (now in its eighth edition)—accessible style, clear examples, and time-tested approaches—but in a streamlined version focusing on critical topics such as copyright, document design, usability, information technologies (including the Internet), and communication in cyberspace. The book retains key qualities of the larger text but in a smaller, concise, technology-centered volume. Students and faculty alike will appreciate its trim size, content, and direct access to information.

Audience for This Book

Most technical communication texts have dual audiences. One audience is instructors, who use textbooks to plan a syllabus, design assignments, and create lectures and discussions. This *Concise Guide* is suitable for a range of instructors, from experienced to novice. All instructors will find this text easy to digest, streamlined in its use of features, and relevant to current technology topics. Novice instructors will find useful examples, exercises, and review checklists; in addition, the companion Web site offers online tutorials, additional exercises, teaching tips, and other teaching resources. Experienced instructors will find that the concise format allows for enhancements within their classrooms without restricting them to one perspective or set of examples.

The student audience for this text is also varied. Students in introductory technical communication will find that the text and Web site contain fundamental concepts, situational strategies, and other supporting features. Advanced students will be able to move quickly into issues of audience, purpose, and design. All students will appreciate the emphasis on the Internet, visual communication, and usability. Examples throughout this text reflect a variety of majors. Students from engineering, science, health care, and other disciplines will find this book useful and relevant, whether in a traditional technical writing class or in a writing-intensive section of their major.

Hallmarks of the First Edition

Layered approach. Instructors can use this *Guide* alone or in combination with their own materials. Instructors may wish to teach the chapters in order, which allows for a logical teaching sequence, especially for an introductory course with students from mixed disciplines. Yet the chapters, and the modules within each chapter, can be taught in virtually any sequence.

Range of skill levels. Students in technical communication courses often span a wide range of writing skills. While this book contains a solid section on grammar and style (Chapter 3 plus Appendix A), certain features of the Web site will be particularly useful for students who need help in this area. The Web site offers intensive exercises, links to style guides, and links to online writing centers.

Compact format. For instructors and students alike, a shorter, more compact text is extremely appealing. Instructors will appreciate the small size, because it allows them to use supplementary material without overloading the student. Students will value a concise text, because material is easy to look up, access, and carry in a backpack.

In addition to its compact size, the *Concise Guide* offers a unique combination of features:

- real examples taken from industry, government, and high technology, reproduced to look like the originals
- sections on copyright, ethics, and social issues
- sections on usability, document design, and page layout from a human factors perspective
- a thorough chapter on visual communication
- a cutting-edge chapter on technical communication in cyberspace
- end of chapter items including review tables, and exercises titled Focus on Writing, The Collaboration Window, and The Global Window. A Click on This section in each chapter provides links to relevant Web resources.

An additional distinguishing feature of this book is its modular, chunked format, allowing the entire book or portions of it to be readily accessed and easily ported into Web-based modules for instructors who wish to teach portions of their course via the Web. The *Concise Guide to Technical Communication* is also available as an electronic book. You can get more information on this option by going on the web site ebookcity.com.

How This Book Is Organized

This book is organized into two parts. Part One, Technical Communication Techniques and Considerations, covers issues of central importance to today's technical communicator: audience, purpose, usability, research, the Internet, ethics, graphics, document design, and copyright. Part Two, Situations and Applications, incorporates the considerations from Part One in treating various types of workplace communication (email, memos, reports, specifications, oral reports, Web pages, and the like).

Instructional Supplements

These ancillary materials are available to accompany *A Concise Guide to Technical Communication*:

- An Instructor's Manual includes teaching tips, chapter quizzes, style exercises, grammar exercises, and other additional individual and collaborative activities.
- A dedicated Web site provides activities that take students and instructors beyond the textbook. The site includes chapter overviews, individual and collaborative projects, and links to resources for both students and instructors. For more information visit www.ablongman.com/gurak.

- The Interactive CD-ROM Edition of *A Concise Guide to Technical Communication* offers the new streamlined format of the text in a multimedia environment. Designed using the Versaware engine, the interactive edition has highlighting, note-taking, and searching capabilities that allow the students to tailor their reading experiences.
- *Daedalus Online*® provides an Internet-based collaborative writing environment for students. The program offers prewriting strategies and prompts, computer-mediated conferencing, peer collaboration and review, comprehensive writing support, and secure, 24-hour availability. For educators, *Daedalus Online* offers a comprehensive suite of tools for managing an online class, dynamically linking assignments, and facilitating a heuristic approach to writing instruction. *Daedalus Online* is available at a discount when bundled with the *Concise Guide.* For more information, visit http://daedalus.pearsoned.com/ or contact your Addison Wesley Longman sales representative.
- *Visual Communication: A Writer's Guide,* by Susan Hilligoss, introduces document design principles that writers can apply across various genres of writing, including academic papers, résumés, business letters, Web pages, brochures, newsletters, and proposals. Emphasizing audience and genre analysis, the guide shows how readers' expectations influence and shape a document's look. Practical discussions of space, type, organization, pattern, graphic elements, and visuals are featured, along with planning worksheets and design samples and exercises.
- *Researching Online,* Fourth Edition, by David Munger and Shireen Campbell, offers a comprehensive introduction to research on the Internet, from email and the World Wide Web, to synchronous discussions and HTML coding.

Acknowledgments

Thank you to the teachers and scholars who reviewed this manuscript: Christine Abbott, Northern Illinois University; Roger Bacon, Northern Arizona University; Marck Beggs, Henderson State University; Lee Brasseur, Illinois State University; Linda Breslin, Texas Tech. University; Eva Brumberger, University of Wyoming; Patricia Cearley, South Plains College; Dave Clark, Iowa State University; Daryl Davis, Northern Michigan University; Ray Dumont, University of Massachusetts; Patrick Ellingham, Broward Community College; Julie Freeman, Indiana University-Purdue University; Lucy Graca, Arapahoe Community College; Kay Harley, Saginaw Valley State University; Joyce Harlow, Rogue Community College; William Wade Harrell, Howard University; Mitchell Jarosz, Delta College; Robert Johnson, Miami University of Ohio; Dan Jones, University of Central Florida; Karla Kitalong, Michigan Technological University; Renee Kupperman, University of

Arizona; Mary Massirer, Baylor University; Robert McEachern, Southern Connecticut University; Brad Mehlenbacher, North Carolina State University; Yvonne Merrill, University of Arizona; Dennis Minor, Louisiana Tech. University; Paul Morris, Pittsburgh State University; Joe Moxley, University of South Florida; Roland Nord, Minnesota State University; Alice Philbin, James Madison University; Carolyn Rude, Texas Tech. University; David Alan Sapp, New Mexico State University; Carol Senf, Georgia Institute of Technology; Katherine Staples, Austin Community College; Tom Stuckert, University of Findlay, Zacharias Thundy, Northern Michigan University; Janice Tovey, East Carolina University; Alex Wang, Normandale Community College; Martin Wood, University of Wisconsin-Eau Claire; Judith Wooten, Kent State University.

Also, thank you to Anne Smith and Arlene Bessenoff for their work with the early visions for this book. Thank you to Rich Wohl, Lynn Huddon, and Carlise Paulson of Longman Publishers, to David Munger of the Davidson Group, and to Janet Nuciforo, for their editorial support of this book. Thank you to Carol Felts for her editorial assistance. Finally, we wish to thank our colleagues and families for their ongoing support.

Laura J. Gurak
John M. Lannon

Technical Communication Techniques and Considerations

chapter 1
An Introduction to Technical Communication

chapter 2
A World of People and Purposes

chapter 3
Designing Usable Information

chapter 4
Performing Research for Technical Communication

chapter 5
Technical Communication in a Digital World

chapter 6
Ethical and Social Issues in Technical Communication

chapter 7
Copyright and Privacy

chapter 8
Graphics and Visual Information

chapter 9
Page Layout and Document Design

CHAPTER 1

An Introduction to Technical Communication

Communicating about technology

Main characteristics of technical communication

Types of technical communication

Technical communication in the workplace

Technical communication in an electronic age

Societal dimensions of technical communication

Ethical dimensions of technical communication

Review checklist

Exercises

The collaborative window

The global window

Click on this

Communicating about technology

We live in a world in which many of our everyday actions depend on complex but important technical information. When you purchase a wallet-sized calculator, for example, the instruction manual is often larger than the calculator itself. When you install any new device, from a VCR to a microwave oven to a cable modem or new computer, it's the setup information that you look for as soon as you open the box. Household appliances, banking systems, online courses, business negotiations, government correspondence and affairs, and almost every other aspect of your daily life is affected by technologies and technical information.

Technical communication has existed since the very earliest times of human writing. The Sumerians, in 3200 BCE, used a stylus and a block of wet clay to record information (Wilford, 1999). But most people trace the rise of technical communication as a profession to the United States after World War II. The rapid development of new technologies during this time created a need for accompanying technical information, such as instructions, manuals, and documentation. And in the 1970s, when the personal computer was invented, well-designed technical communication became vital as more and more nontechnical people began using computers, software, and other devices. Today, with a large percentage of the population using the Internet, banking via the telephone, and interacting with technology in so many other ways, we all recognize the importance of well-designed technical information.

Main characteristics of technical communication

Technical communication is the art and science of making complex technical information *accessible, usable,* and *relevant* to a variety of people in a variety of settings. To some extent, effective technical communication is an art, because it requires an instinct for clear writing and good visual design. More importantly though, technical communication is also a science, a systematic process that involves certain key principles and guidelines. The following principles characterize effective technical communication.

Accessibility

Information is accessible if people actually can get to it and understand it. If documentation for a help system is included on CD-ROM, the people using this information must access a CD-ROM drive in order to use the information. If a set of instructions is being distributed across the globe, these instructions must be written in various languages in order to be accessible to international users.

A group of technical editors at IBM has developed a list of "quality characteristics," which help them determine if their technical documentation meets high standards and is of superior quality. These characteristics suggest specific ways in which communication can be made accessible:

- Accuracy—has no mistakes or errors
- Clarity—avoids ambiguity
- Completeness—includes all necessary information
- Concreteness—uses concrete examples and language
- Organization—follows sequences that make sense for the situation
- Visual effectiveness—uses layout, screen design, color, and other graphical elements effectively (Hargis, Hernandez, Hughes, & Ramaker, 1997, p. 2)

Usability

Usable information is more *efficient* for your audience, because it allows readers to perform the task or retrieve the information they need. Usability is often measured by studying the design of the table of contents, index, headings, and page layout, as well as determining if the language is written at the appropriate technical level. When technical communicators assess a document's usability, they may want to know how long it took a person using the document to find specific information and whether this information could be located using the index or table of contents. For instance, a manager may consult the company's Employee Handbook for information about vacation time. If the manager cannot find this information and cannot do so quickly, the document would not be considered usable and would need to be revised. (Learn more about usability in Chapter 3.)

Relevance

Relevant information maintains a focus on the specific *audience*—the readers, listeners, viewers—who need information, not piles of useless data. Information is relevant if the audience can apply it to the task at hand. For instance, if a person is interested in how to use Internet service provider (ISP) software to connect to the Internet, the documentation should explain how to install the software and dial up the ISP and not digress into a history of how the Internet developed. Or, for an audience of general computer users who want to install a sound card, overly technical language is inappropriate. Relevant information also maintains a focus on the *purpose* of the communication. Although the history of sound cards might be interesting to some engineers, the purpose of the communication (how to install the sound card) dictates that this history is not relevant.

Often, technical communication is thought of in relation to the documents and technologies described above; that is, as communication designed to teach a general audience how to perform a specific task involving a common sort of technology—how to set up a VCR, install a new sound card in a PC, or install the mulching blade on a lawn mower. But technical information is also used by technical specialists, managers, and others. A surgeon performing heart surgery must have clear information about how to install a pacemaker. A government research scientist must have accurate instructions about how to write a grant or how to perform a particular experiment. An engineer must have access to the right specifications for designing a bridge or configuring an application. In all settings in which people must understand complex information, there is a need for technical communication.

Consider the following example, which illustrates just one of these ideas: how technical information is made *accessible* by the writer's use of consistent terminology. Unlike some forms of writing, in which authors are often told to vary their choice of words, technical communication strives for accuracy by using consistent terminology when referring to the same item or task.

For example, assume that you have just purchased a new children's toy that needs assembling. In one case, the writer of the instructions decided that she would vary the terminology. Here is her first draft:

Your new RetroRocket comes with the following parts:

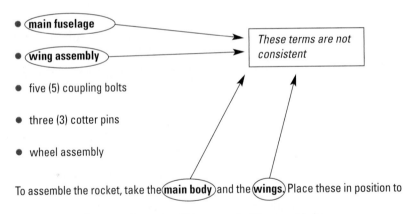

To assemble the rocket, take the (main body) and the (wings) Place these in position to each other as shown on the carton. Take several of the round bolts . . .

The list of parts says "main fuselage" and "wing assembly" but the instructions say "main body" and "wings". Are these the same pieces? If you don't use the same terms, how will your reader know?

No list of technical communication characteristics can address all possible situations and decisions you will encounter. Wherever you work, you will face challenges that ultimately require your best analytical and social skills. For instance, "clarity" is a typical characteristic of technical communication. But in some cases, being crystal clear may be impossible, because you can't get enough information. Or, clarity and directness may be seen as overly blunt and offensive to some readers. As you read this book, remember that communication ultimately takes place in companies, among people, in situations that change, and often under time and budget pressures.

Types of technical communication

Following are some common forms of technical communication. Although these categories can overlap considerably, they should give you a feel for the kinds of documents technical communicators produce.

- **Manuals.** Almost every technology product or service is accompanied by a manual. Manuals may include information on how to use a product, along with background information, such as technical specifications or lists of materials. You certainly have used such manuals—to connect the components of your sound system, to do routine maintenance on your bicycle or inline skates, or to set up your answering machine.
- **Procedures.** Procedures are an important form of technical communication. Procedures explain how to perform a task or how a particular process happens. Many companies maintain standard operating procedures ("SOPs") for tasks such as how to test soil samples or how to access corporate databases.
- **Instructions.** Instructions resemble manuals and procedures in that they explain how to do something. However, instructions are often very specific, systematic lists of the actual steps involved in using a product or performing a procedure. For instance, if you purchase a memory upgrade for your computer, you will probably receive a list of instructions on how to install this upgrade. This list may be a separate document or part of a manual or larger set of instructions.
- **Quick reference cards.** In some situations, a long list of procedures or instructions is inappropriate, because the user is already familiar with the "big picture." For instance, you may regularly call home to access your voice mail, and you may have the primary commands memorized. But there are certain tasks you may perform infrequently, such as changing your outgoing message from another phone. For tasks that users perform on a limited basis, a short summary of the keypad commands may be all that is needed. These commands can often fit on a quick reference card designed to fit in a wallet or fit in with the actual device (over the telephone keypad, for instance).
- **Reports.** There are many types of reports, including recommendation reports and analytical reports. Reports generally focus on a specific problem, issue, or topic. They may recommend a course of action or analyze a particular technology or situation. For example, a task force in your com-

munity may be studying plans for highway expansion or a new shopping center. After completing an initial study, task forces often present reports to the city council or other decision makers, and written copies of these reports are available for public review.

- **Proposals.** Proposals make specific recommendations and propose solutions to technical problems. A proposal's purpose is usually to persuade readers to improve conditions, accept a service or product, or otherwise support a plan of action. Proposals are sometimes written in response to calls for proposals (CFPs) or requests for proposals (RFPs). For example, a nonprofit childcare facility may seek safer playground equipment, or a pharmaceutical company may wish to develop a new Web-based education program for its employees. These organizations would issue RFPs, and each interested vendor would prepare a proposal that examines the problem, presents a solution, and defines the process and fees associated with implementing the solution.
- **Memos.** A vital form of technical communication, memos serve various purposes: to inform, to persuade, to document, or to encourage discussion. Memos are usually brief and follow a format that includes a header ("to," "from," "date," "re") and 1–2 pages of body text. An employee might write a memo to his manager requesting a pay raise; an engineer might write a memo to her design team explaining a technical problem and offering a solution; a team of students might write a memo to their instructor explaining their progress on a class project.
- **Email.** Email is, essentially, the electronic version of a memo. In fact, most email is patterned after the memo, with a header containing fields for "to," "from," "date," and "re" already built in. Yet email messages are more pervasive than paper memos. In most work settings, people use email to relay scheduling, policy, procedure, and miscellaneous information. They communicate via email with clients, customers, and suppliers—as well as with associates worldwide. People are more inclined to forward email messages, and tend to be more casual and write more hastily than they would with paper memos.

Although these forms are common, many others exist, depending on the company or profession. Nursing, for example, requires specific forms for documenting a patient's medical condition; engineering has its own types of technical communication. In addition, the specific audience and purpose in each situation will determine the appropriate type of communication.

Various types of communication can also be formatted and packaged in various media:

- CD-ROM
- Internet Web pages (the entire worldwide Internet)
- Intranet Web pages (an internal network)
- Electronic text, including email or attachments
- Online help

- Printed matter, including books, paper memos, bound reports, and brochures
- Training sessions or oral presentations

Technical communication in the workplace

People who make technical information accessible to different audiences are called "technical communicators." In more and more organizations, this position is a full-time job, with titles including the following:

- Technical writer
- Technical editor
- Web designer
- Online documentation specialist
- Information developer
- Instructional developer

Technical communicators write and design documentation, online information, software interfaces, and other documents and materials for users of high technology. Technical communicators also write technical memos, reports, grant applications, and other specialized documents.

Virtually all technical professionals, at one time or another, function as part-time technical communicators. These technical experts are often required to present their knowledge to nonexpert audiences. For instance, a nuclear engineer testifying before Congress would need to explain nuclear science to nonscientists and to address the concerns of policymakers. People from many other walks of life (lawyers, health care professionals, historians, managers, and so on) communicate specialized information to nonexpert audiences:

- Medical professionals discuss health matters with patients.
- Attorneys interpret the law for clients.
- Historians describe complex historical events for people who did not experience those events.
- Managers interpret business objectives for those they supervise.

Technical communication in an electronic age

Electronic technologies allow far more communication today than ever before. Long before people had a telephone, voice mail, email, Web access, a pager, or a cell phone, they communicated by speaking and writing. We still do this: we

Figure 1.1 ■ Electronic communication devices. We live in an age of information overload. Effective technical communication can help people manage and sort through this information.

have meetings or go down the hall to speak with a colleague, and we still send letters, print newspapers, and read books. But now we send and receive information through an even greater number of channels. And while we may enjoy the efficiency of email, the convenience of voice mail, and the cost savings of teleconferences, it is also apparent that many professionals are suddenly struggling with information overload (Figure 1.1). And the more information people receive on a daily basis, the more urgent the need to make sure this information is accessible, usable, and relevant. In short, our information-saturated society cries out for effective technical communication.

Societal dimensions of technical communication

Good technical communication has a societal component, because it can make important topics in science and technology (e.g., genetically modified organisms, cloning, computers that diagnose disease) understandable to the general public. Such communication opens doors to new information—doors that might otherwise remain shut if the information were hard to read, too technical, or impossible to interpret. If the general public tried to learn about these topics by reading technical journals, they would come away scratching their heads, because the language and presentation would be too technical for general readers. But if this information is written to match the reader's level of knowledge, readers can understand these important topics. In a world in which science and technology play major roles in our everyday lives, technical communication becomes increasingly important. When you create effective technical communication, you not only help others use the information, but you also help people learn about important ideas.

Ethical dimensions of technical communication

Technical communication involves an ethical stance as well, because the words, fonts, graphics, and colors that convey the information may influence your audience's perception, interpretation, and understanding. For example, think of the many advertising claims hinting that certain herbal remedies may cure diseases. These claims, technical in nature, often have no basis in traditional scientific methods. Yet some technical communicator chose (or was instructed) to write these words. The workplace pressures of communicating what the boss wants, or what will make more money for the company, are often at odds with the ethical pressures to present information fairly and accurately. Visual communication, such as charts and graphs, can also be misused. Later chapters address the ethical issues involved in technical communication. In the end, you will need to balance your own ethical stance against the interests of others, including your company and your customers or end users.

 Review Checklist

Characteristics of Quality Technical Communication	*Questions You Should Ask*
Accessible—users can find what they need	Is the information *accurate*?
	Is the language *clear* and unambiguous?
	Is the information *complete*?
	Are the examples *concrete*?
	Is the material appropriately *organized*?
	Is *visual information* (layout, screen design, color) used effectively?
Usable—users can use the information to perform a task	Can users find what they need in an *efficient* manner?
	Is language at an appropriate *technical level*?
	Does the document contain a *table of contents*, index, or other such device?
Relevant—users can relate the content to their task or project	Is the material appropriate for this *audience*?
	Is the material appropriate for and relevant to the *purpose* at hand?

EXERCISES

1. Locate an example of a technical document and bring it to class. Use the review chart above to explain to other students why your selection can be called "technical communication." Explain how your selection is accurate, visually effective, usable, and focused on the audience.

2. Consider how many types of electronic communication (telephone, voice mail, email, fax, pager, and so on) you are involved with daily. List these technologies according to how useful each might or might not be for a specific audience or purpose. For example, if you wanted to ask your boss for a raise, would you use email, the phone, or a face-to-face meeting? If you wanted to explain a technical concept, would a Web page be appropriate? (Think up a few other scenarios of your own.) Discuss your answers with the entire class.

3. Focus on ⟩⟩⟩Writing. Assume that a friend in your major thinks that technical communication skills are not needed—that anyone can write or design information without thought to issues of access, usability, or relevance. Write your friend a memo based on the information in this chapter explaining why you think these assumptions are mistaken. Use examples (brochures, Web pages, other technical communication) to support your position. (See pages 187–189 for details on memo formats.)

The Collaboration Window

Most writing and communicating, especially in the workplace, is done collaboratively: that is, it is done by and among many people and takes numerous ideas and suggestions into account. In class, form teams of students who have the same or similar majors or interests.

Your assignment. Create a list of technical terms and concepts, with short explanations, that you feel are important for people to understand your major or career interest; in other words, create a miniature "dictionary" for your major or field. Your list may consist of only 10 terms and must fit on a single page. Collaborate on forming this list as follows:

- Each person in the group should create an individual list.
- When everyone is done, compile these lists into one master list. You will need to negotiate among members of your group about what 10 terms to keep and how to define these terms. Share your list with the other groups in class.

The Global Window

Technical communication is an international activity. Technical products and services are used around the world, and communicators need to create information that is attentive to international needs. For example, if a company is shipping microcassette tape recorders to several countries, the documentation must be written in clear English that can easily be translated and contains internationally recognized symbols or visual information.

Your assignment. Locate technical documentation that is written in English plus several other languages (instructions for household appliances, tools, or stereo equipment is often written in several languages). How many languages were used? Why did this company select these languages? Compare your findings with those of other students.

To learn more about global communication, use a Web search engine to locate information about the International Standards Organization (ISO). This group specializes in creating technical and communication standards for worldwide use. Identify a particular aspect of this site that you find interesting or that is related to your major, and share this information in class.

Click on This

Locate an example of technical communication on the Web. Try a few of the Web pages listed below to get you started, or use a search engine to locate Web pages related to your career or technical interest. Bring printouts of one or several pages to class, and work with a group of students to identify examples—both good and bad—of the characteristics (accessible, usable, relevant, and so on) discussed in this chapter.

- www.nlm.nih.gov/locatorplus
  The National Library of Medicine's Web catalog of over 5.3 million books and other materials.

- www.elib.cs.berkeley.edu/photos/flora
  The University of California, Berkeley, Digital Photo Project of more than 20,000 color images of native and naturalized California plant species and habitats.

- www.lanl.gov/worldview
  The home page for the Los Alamos National Laboratory, with photos and descriptions of the lab's research projects.

CHAPTER 2

A World of People and Purposes

People and purposes

Analyzing your audience

Analyzing the communication purpose

Analyzing the communication context

Conducting an audience/purpose interview

More tools for understanding audience

Using information from your analysis

Audiences are not passive

Typical audiences and purposes for technical communication

Review checklist

Exercises

The collaboration window

The global window

Click on this

People and purposes

All forms of technical communication are ultimately intended for an *audience:* the readers, listeners, viewers, and users who need information so they can make decisions or perform tasks. A good technical communicator always designs information with an audience in mind, carefully reviewing a vast array of information, selecting what is important, and crafting the information into a useful tool for a specific group of people.

For example, the brochure in Figure 2.1 is designed by a biomedical device company to inform a specific audience: physicians and other health care professionals who treat patients with heart conditions. Knowing how this audience of doctors and nurses feels about meeting individual patient needs, the writers begin with the following sentence: "The Medtronic Kappa™ Generation of Pacing Systems offers greater choice in patient management because no two patients are exactly the same." The writers then outline several key features of this product.

Audience is an important consideration in all kinds of writing but especially in technical writing, which is far more *user-centered* than other writing. When you write a poem or an essay, for example, you often express your personal feelings and thoughts on the subject. But as a technical writer, your first concern is to provide information the audience *needs*. This is not to say that technical communication involves no thinking and feeling, only that the technical writer's thoughts and feelings must take a back seat to the audience's needs.

According to technical communication expert R. Johnson (1997), you can never know every member of your audience, so you always have to do some guessing about their needs. However, you should interact with and seek feedback from actual audience members as much as possible.

All forms of technical communication are also intended for specific *purposes:* the workplace settings, situations, and reasons for a particular form of communication. If the purpose is to persuade, this will influence the form of communication. If the purpose is to inform, this will affect the language, format, and other features of the communication. Many documents have multiple purposes. For example, the primary purpose in most instruction manuals is to teach an audience how to use the product. But for ethical and legal reasons, companies are also concerned that people use the product safely. An instruction manual for a cordless drill (Figure 2.2) for example, begins with a page of safety instructions.

Any simple message can be conveyed in many different ways, depending on how it is constructed for different audiences. Information about a new cancer treatment may appear in a medical journal for health care professionals, a textbook for nursing or medical students, and a newspaper article for general audiences. In each case, the writer would need to consider very different audience and purpose features. An audience of medical professionals will

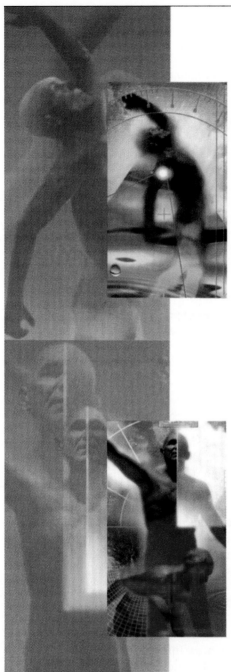

Medtronic Kappa™ Generation of Pacing Systems

The Medtronic.Kappa™ Generation of Pacing Systems offers greater choice in patient management because no two patients are exactly the same.

The 700 Choice -
Adapts to Your Patient, As You Would Want

Advanced, second generation mode switch

Extensive adaptive capabilities, including Capture Management™ and Sensing Assurance™

Patient-specific ℓ. Rate Response with single sensor

Automated follow-up efficiencies

The 400 Choice -
Expanding Rate Response Therapy

Patient-specific ℓ. Rate Response with integrated sensors

Automatic diagnostics enhancing heart rate therapy

FIGURE 2.1 ■ A technical brochure. This brochure about a heart pacemaker system is designed to speak to a specific audience.
Source: Medtronic Kappa™ Generation of Pacing Systems, Medtronic, Inc.

IMPORTANT SAFETY INSTRUCTIONS

WARNING: When using Electric Tools, [always follow] basic safety precautions to reduce risk of fire, electric shock, and personal injury, including the following:

READ ALL INSTRUCTIONS

1. **KEEP WORK AREA CLEAN.** Cluttered areas and benches invite injuries.
2. **CONSIDER WORK AREA ENVIRONMENT.** Don't expose power tools to rain. Don't use power tools in damp or wet locations. Keep work area well lit.
3. **GUARD AGAINST ELECTRIC SHOCK.** Prevent body contact with grounded surfaces. For example: pipes, radiators, ranges, refrigerator enclosures.
4. **KEEP CHILDREN AWAY.** All visitors should be kept away from work area. Do not let visitors contact tool or extension cord.
5. **STORE IDLE TOOLS.** When not in use, tools should be stored in a dry, and high or locked-up place - out of reach of children.
6. **DON'T FORCE TOOL.** It will do the job better and safer at the rate for which it was intended.
7. **USE RIGHT TOOL.** Don't force small tool or attachment to do the job of a heavy-duty tool. Don't use tool for purpose not intended. For example, don't use a circular saw for cutting tree limbs or logs.
8. **DRESS PROPERLY.** Do not wear loose clothing or jewelry. They can be caught in moving parts. Rubber gloves and non-skid footwear are recommended when working outdoors. Wear protective hair covering to contain long hair.
9. **USE SAFETY GLASSES.** Also use face or dustmask if operation is dusty.
10. **DON'T ABUSE CORD.** Never carry tool by cord or yank it to disconnect from receptacle. Keep cord from heat, oil, and sharp edges.
11. **SECURE WORK.** Use clamps or a vise to hold work. It's safer than using your hand and it frees both hands to operate tool.
12. **DON'T OVERREACH.** Keep proper footing and balance at all times.
13. **MAINTAIN TOOLS WITH CARE.** Keep tools sharp and clean for better and safe performance. Follow instructions for lubricating and changing accessories. Inspect tool cords periodically and if damaged have repaired by authorized service facility. Inspect extension cords periodically and replace if damaged. Keep handles dry, clean, and free from oil and grease.
14. **DISCONNECT TOOLS.** When not in use, before servicing, and when changing accessories, such as blades, bits, cutters.
15. **REMOVE ADJUSTING KEYS AND WRENCHES.** Form habit of checking to see that keys and adjusting wrenches are removed from tool before turning it on.
16. **AVOID UNINTENTIONAL STARTING.** Don't carry plugged-in tool with finger on switch. Be sure switch is off when plugging in.
17. **OUTDOOR USE EXTENSION CORDS.** When tool is used outdoors, use only extension cords intended for use outdoors and so marked. (See page 4 for more information about extension cords.)
18. **STAY ALERT.** Watch what you are doing. Use common sense. Do not operate tool when you are tired.
19. **CHECK DAMAGED PARTS.** Before further use of the tool, a guard or other part that is damaged should be carefully checked to determine that it will operate properly and perform its intended function. Check for alignment of moving parts, binding of moving parts, breakage of parts, mounting, and any other conditions that may affect its operation. A guard or other part that is defective should be properly repaired or replaced by an authorized service center unless otherwise indicated elsewhere in this instruction manual. Have defective switches replaced by authorized service center. Do not use tool if switch does not turn it on and off.
20. **DO NOT OPERATE** portable electric tools near flammable liquids or in gaseous or explosive atmospheres. Motors in these tools normally spark, and the sparks might ignite fumes.

CAUTION: When drilling into walls, floors, or wherever "live" electrical wires may be encountered, DO NOT TOUCH THE CHUCK! Hold the drill only by the plastic handle to prevent electric shock if you drill into a "live" wire.

We understand that safety rules make some pretty dry reading, but they really are important. If you just skimmed them, please go back and thoroughly read them. Thank you.

SAVE THESE INSTRUCTIONS

Figure 2.2 ■ **Safety instructions for operating a cordless drill.** This cordless drill manual begins with a page of safety instructions.
Source: Black & Decker Instruction Manual, © 1993. Black & Decker (U.S.) Inc.

understand technical terms, but general readers will not. In terms of purpose, medical students are reading so they can apply the information, while general audiences are often reading for nonspecific learning. Thus, each of these articles will differ in its language, content, organization, illustrations, and overall design. The more you understand about your audience and purpose, the more your communication will meet user needs.

Figures 2.3 and 2.4 show two pieces of information, both about the prescription allergy medication Claritin. The Web site is designed for a general audience of patients who have questions or want more information about this medicine. The page from the *Physician's Desk Reference* (PDR) is also designed to answer questions and provide information but for an audience of physicians and health care professionals, not patients. Each item addresses the same topic but is designed and written for very different audiences.

Of course, decisions about audience are not always as straightforward as those in Figures 2.3 and 2.4. For instance, an engineer might write a memo to a colleague in highly technical terms that only a fellow engineer would clearly understand. Later this same memo might be forwarded to a manager in another department who has less technical understanding of the topic at hand, and this manager might need to base an important decision on the information in the memo. In this case, the original writer should have attached a sheet of definitions or background information. Never assume that you know your audience with absolute certainty, and never stop questioning how to meet your audience's needs more effectively.

Analyzing your audience

In preparing a technical communication product, you generally begin by analyzing your audience through a series of questions like these:

- Who will be reading/listening to/using this material?
- What special characteristics do they have?
- What is their background and attitude toward the subject?

Most people already know more than they think they do about analyzing an audience. Imagine that you are asked to give a presentation on global warming to a group of school children. Later, you are asked to speak on the same subject to a group of manufacturing executives. In preparing to speak to the children, you would probably think of ways to make the topic understandable: for instance, using simple language and comparing your ideas to things familiar to the children. In preparing to speak to the manufacturing executives, you would change your approach, using more technical terms and referring to topics they care about, such as the effects of global warming on their industries.

In short, you would have performed a rudimentary *audience analysis* by assessing the characteristics and interests of the two different audiences, then reshaping the information to fit what you know about each group. Yet it is best to be more systematic about analyzing an audience, because a communicator's assumptions are sometimes wrong. For example, you might assume that the manufacturing executives know quite a bit about global warming and therefore use technical terms or refer to complex concepts. But what if your assumption is wrong? What if these executives actually know very little about the subject? Instead of relying on your assumption, learn all you can about this audience's background knowledge before the presentation.

Most communication situations have an immediate audience. This is your *primary audience*. For instance, a set of instructions for installing new email software for the office might be directed primarily at computer support staff. But most documents also have *secondary audiences,* people outside the circle of those who need the information urgently. A secondary audience for software instructions might be managers, who will check the instructions for company policy, or lawyers, who will make sure the instructions meet various legal standards.

ANAlyziNG THE COMMUNiCATiON puRpOSE

As you analyze your audience, you also need to consider the purpose of your message by asking questions like these:

- Why is this communication important?
- Why is it needed?
- What will users do with this information?

People use technical information for various purposes: to perform a task, learn more about a subject, or make a decision. If the communicator has one purpose in mind when preparing the information but the audience has a different purpose, the message will be useless.

For example, you may have encountered Web sites where the purpose of the page seems at odds with your purpose for visiting the site. Let's say you hear about a new Web site that sells books about bird-watching in South America. As an avid bird-watcher about to travel to Brazil, you decide to check out this site. When you first connect, you are impressed with the bright colors and cute bird sounds. And as you click on each book selection, you enjoy an array of birds that come flying across your screen. Yet you cannot locate any descriptive

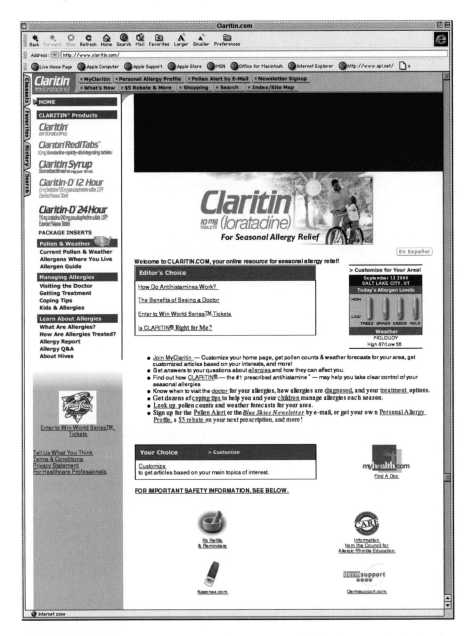

FIGURE 2.3 ■ Claritin Web site. This Web site of medical information is designed for a more general audience.

Source: Claritin Web site (www.claritin.com).

CLARITIN-D® 24 HOUR   ℞
brand of loratadine and
pseudoephedrine sulfate, USP
Extended Release Tablets

DESCRIPTION

CLARITIN-D® 24 HOUR (loratadine and pseudoephedrine sulfate, USP) Extended Release Tablets contain 10 mg loratadine in the tablet coating for immediate release and 240 mg pseudoephedrine sulfate, USP in the tablet core which is released slowly allowing for once-daily administration.

Loratadine is a long-acting antihistamine having the empirical formula $C_{22}H_{23}CIN_2O_2$; the chemical name ethyl 4-(8-chloro-5,6-dihydro-11H-benzo[5,6]cyclohepta[1,2-b] pyridin-11-ylidene)-l-piperidinecarboxylate; and the following chemical structure:

The molecular weight of loratadine is 382.89. It is a white to off-white powder, not soluble in water, but very soluble in acetone, alcohol, and chloroform.

Pseudoephedrine sulfate is the synthetic salt of one of the naturally occurring dextrorotatory diastereomers of ephedrine and is classified as an indirect sympathomimetic amine. The empirical formula for pseudoephedrine sulfate is $(C_{10}H_{15}NO)_2 \cdot H_2SO_4$; the chemical name is α-[1-(methylamino) ethyl]-[S-(R^*, R^*)]-benzenemethanol sulfate (2:1)(salt); and the chemical structure is:

The molecular weight of pseudoephedrine sulfate is 428.54. It is a white powder, freely soluble in water and methanol and sparingly soluble in chloroform.

The inactive ingredients for oval, biconvex CLARITIN-D 24 HOUR Extended Release Tablets are calcium phosphate, carnauba wax, ethylcellulose, hydroxypropyl methylcellulose, magnesium stearate, polyethylene glycol, povidone, silicon dioxide, sugar, titanium dioxide, and white wax.

CLINICAL PHARMACOLOGY

The following information is based upon studies of loratadine alone or pseudoephedrine alone, except as indicated.

Loratadine is a long-acting tricyclic antihistamine with selective peripheral histamine H_1-receptor antagonistic activity.

Human histamine skin wheal studies following single and repeated oral doses of loratadine have shown that the drug exhibits an antihistaminic effect beginning within 1 to 3 hours, reaching a maximum at 8 to 12 hours, and lasting in excess of 24 hours. There was no evidence of tolerance to this effect developing after 28 days of dosing with loratadine.

Pharmacokinetic studies following single and multiple oral doses of loratadine in 115 volunteers showed that loratadine is rapidly absorbed and extensively metabolized to an active metabolite (descarboethoxyloratadine). Approximately 80% of the total dose administered can be found equally distributed between urine and feces in the form of metabolic products after 10 days. The mean elimination half-lives found in studies in normal adult subjects (n = 54) were 8.4 hours (range = 3 to 20 hours) for loratadine and 28 hours (range = 8.8 to 92 hours) for the major active metabolite (descarboethoxyloratadine). In nearly all patients, exposure (AUC) to the metabolite is greater than exposure to parent loratadine. Loratadine and descarboethoxyloratadine reached steady state in most patients by approximately the fifth dosing day. The pharmacokinetics of loratadine and descarboethexyloratadine are dose independent over the dose range of 10 to 40 mg and are not significantly altered by the duration of treatment.

In vitro studies with human liver microsomes indicate that loratadine is metabolized to descarboethoxyloratadine predominantly by P450 CYP3A4 and, to a lesser extent, by P450 CYP2D6. In the presence of a CYP3A4 inhibitor ketoconazole, loratadine is metabolized to descarboethoxyloratadine predominantly by CYP2D6. Concurrent administration of loratadine with either ketoconazole, erythromycin (both CYP3A4 inhibitors), or cimetidine (CYP2D6 and CYP3A4 inhibitor) to healthy volunteers was associated with significantly increased plasma concentrations of loratadine (see Drug Interactions section).

In a study involving 12 healthy geriatric subjects (66 to 78 years old), the AUC and peak plasma levels (C_{max}) of both loratadine and descarboethoxyloratadine were significantly higher (approximately 50% increased) than in studies of younger subjects. The mean elimination half-lives for the elderly subjects were 18.2 hours (range = 6.7 to 37 hours) for loratadine and 17.5 hours (range = 11 to 36 hours) for the active metabolite.

In patients with chronic renal impairment (creatinine clearance ≤30 mL/min) both the AUC and peak plasma levels (C_{max}) increased on average by approximately 73% for loratadine; and approximately by 120% for descarboethoxyloratadine, compared to individuals with normal renal function. The mean elimination half-lives of loratadine (7.6 hours) and descarboethoxyloratadine (23.9 hours) were not significantly different from that observed in normal subjects. He-

Continued on next page

Information on Schering products appearing on these pages is effective as of August 15,1999.

Consult 2 0 0 0 PDR® supplements and future editions for revisions

Fɪɢᴜʀᴇ 2.4 ■ **Claritin information from the *Physician's Desk Reference* (PDR).**
This information is designed for a specialized audience of medical professionals.
Source: Physician's Desk Reference (1999), p. 2785.

information explaining how one book differs from another, and you are not sure whether the prices displayed include shipping. Also, you can't find an order form. It appears that your purpose, to locate and perhaps buy a book, conflicts with the purpose of the page, which appears to be more of a fancy digital advertisement than a place where customers can find information and make a purchase.

Just as there are primary and secondary audiences, there is also more than one level of purpose. The primary purpose of a set of instructions for a new bicycle rack might be to help users assemble the rack, but a secondary purpose might be to meet the company's legal obligation to list all parts and inform users about potential hazards.

Analyzing the communication context

Along with audience and purpose, it is also important to understand the context in which the document will be used. Context is related to purpose, but it suggests a slightly different set of questions, such as:

- What are the organizational settings in which the document will be used? For example, will the document be used in training sessions? As part of overall policy documents? As a Web-based customer support site?
- Are there legal issues to consider? For example, are you using material from another source, and if so, do you need to request permission? Are you discussing company projects that may be confidential?
- How much time do people at this company or with this job title have to perform a task? For example, a service technician out in the field may have very little time to locate an answer, but a researcher working on a long-term experiment may have more time to mull over the theoretical points about a topic.
- Are the readers of this document associated with a larger group of professionals (nurses, scientists, teachers), and if so, what professional values might they bring to the situation? For example, medical professionals value the health and life of the patient above all else.
- Are audience members from one culture only, or is this information directed at a cross-cultural audience? Remember that even the United States contains many diverse cultures: not everyone in the United States speaks English as their first language, for example.

These and other issues will affect every choice you make when writing and designing technical communication.

The following chart summarizes important questions about audience, purpose, and context and provides a template worksheet for your own analysis. Modify this chart to suit your specific situations.

Audience	Specific Features
Demographic information	Age, gender ratio, education level, ethnicity
Primary audience	Names, job titles
Secondary audiences	Names, job titles
Audience attitudes toward information	Fearful, receptive, interested, other
Technical understanding of topic	Very knowledgeable, some knowledge, complete novice
Other experience with topic	
Purpose	
Primary purpose	
to learn	List what they want to learn
to obtain background information	List why they need this
to make a decision	List the decisions they will make
to perform a task	List tasks: to build, to design, to install, etc.
Secondary purpose(s)	Legal, marketing, other
Context	
Role within the organization	Managers, engineers, etc.
Political or social situation	Power, decision making
Legal issues	Copyright, patents
Cultural considerations	Cross-cultural audiences
Professional values or affiliations	Engineers, teachers, nurses
Other contextual issues	Due dates, other constraints

Conducting an Audience/Purpose Interview

Some documents have wide audiences: Manuals for household appliances, for example, are sent out to countless users worldwide. You may be able to interview only a few people, but if they represent the average reader or listener, you will have a good sense of how to proceed.

The first step, then, in analyzing your audience is to identify the people available for an interview. Try to interview people from various segments of your audience. Depending on your situation, you may interview people individually or in focus groups (small groups of people brought together for this purpose).

Before the interview, explain what you are doing and ask individuals for an appropriate time to meet or call. Email is a good way to make this initial contact, allowing recipients to respond at their convenience. If you can't set up a face-to-face or phone interview, consider using email to conduct your audience analysis.

But don't send out your analysis questions until your respondent has agreed to participate. (See Chapter 4 for more information on conducting interviews.)

During the interview, cover all the items on your audience analysis worksheet. Pay particularly close attention to the following items:

- **Levels of technical understanding.** How much technical knowledge does the reader have? Will technical terms be familiar or confusing? How much background will be needed to help explain concepts?
- **International issues.** Are audience members from one culture or country or from several countries and cultures? How can the document be written and designed so it is accessible to everyone?
- **Workplace culture or hierarchy.** In what workplace setting will the document be used? Is there a certain style or tone appropriate to this company? Will all levels of employees be using this material, or is it designed for just one or two groups? Does the company have its own style manual?
- **Gender.** What is the gender mix in this audience? How can the document be written and designed so it is fair to both men and women? For example, if a document needs to refer to a job title, use gender-neutral language ("mail carrier" rather than "mailman").
- **Mixed audiences.** In preparing a set of procedures, you would write one way for experienced users (people who've performed this or similar procedures before) and a different way for inexperienced users. But if your audience consists of both groups, you need to include different levels of information within one document: some background and explanation of technical terms for the inexperienced users and some technical terms and concepts for those with experience. You can also use an approach called "layered communication," discussed in Chapter 3.

More tools for understanding audience

Enhance your audience analysis by seeking out other information, such as the following:

- **Corporate style guides.** Companies often publish style guides with their own rules and guidelines for corporate communication. These guides offer specific information on everything from grammar and punctuation to tone and style. A company style guide often describes the audiences for its products.
- **User preference documents.** Many manufacturing and software organizations create documents that assess user preference. These documents are often created after detailed interviews with real customers.
- **Marketing surveys and focus groups.** Marketing departments spend a great deal of time with customers and have a wealth of information to share with you on customer attitudes, preferences, educational levels, and so on.

Using information from your analysis

A thorough audience and purpose analysis will help you make the following decisions as you prepare your document or presentation.

- **Word choice.** An audience's technical level will govern what kind of language you use. A group of software engineers will understand technical language about computing ("remote analog loopback"), but a mixed audience of managers and supervisors may require less technical language and clear definitions of any technical terms you do use. Novices may need nontechnical, reassuring language.
- **Examples.** Good examples can make a technical concept clear and easy to visualize. A document describing how a pacemaker works might compare the pacemaker's action to a more familiar concept, such as a ticking clock.
- **Document format.** An audience often expects a document to conform to a familiar format. In some companies, the standard format for communicating new product information might be a special type of company memo, a prepared form, or an email message with an attachment.
- **Length.** Some audiences, such as busy executives, have no time to read an entire report. In these cases, the report is preceded by an informational abstract that summarizes key information and conclusions. Length is also important in presentations. Some meetings with a busy agenda often limit individual presentations to 10 or 15 minutes.
- **Document genre.** Is your document meant to persuade or inform? Although all documents are implicitly persuasive (in that you want readers to appreciate the quality of your message), some are expected to be explicitly persuasive as well. A sales proposal, for example, explicitly attempts to persuade its audience to purchase the product or service; on the other hand, a research report is usually intended merely to describe the project and interpret the findings.
- **Information you will include.** Make sure the information you include in the document is interesting and useful for your audience. Consider carefully what to leave in and what to leave out based on what you know about audience and purpose. As one expert noted, a product "might have a powerful new help system, but information about [it] is of little interest to the person who is reading about the installation" (Hargis, 1997, p. 13).

Audiences are not passive

Audiences are not merely passive recipients of information. When people read a manual, listen to a presentation, or interact on a Web site, they constantly form opinions of the material, learn new information, and consider new points of view. If they find the information difficult to use, not credible, or insulting, they reject it. So as you analyze your audience, remember that the communication process works both ways. People will react to your ideas in ways that you may not anticipate. Keep an open mind, and be ready to modify your original ideas based on how your audience reacts.

Typical audiences and purposes for technical communication

The following categories of audiences and purposes are presented to give you an overview of how to think about different groups of people in different communication contexts. Obviously, these categories are not exclusive. We all shift in and out of various audience roles: for example, at work you may be a nurse or an engineer, but when you go home, you may function more as a member of the general public. Also, in creating categories, we run the risk of stereotyping. It's impossible to speak about *all* engineers or *all* musicians. Yet it is helpful to consider audience types, because a specific type of audience generally shares a specific type of concern. Take these concerns into consideration as you plan and design your document or presentation.

Scientists search for knowledge to "understand the world as it is" (Petroski, 1996, p. 2). Scientists look for at least 95 percent probability that chance played no role in a particular outcome. They want to know how well a study was designed and conducted and whether the findings can be replicated. Scientists know that their answers are never "final," but open-ended and ongoing: What seems probable today might be rendered improbable by tomorrow's research.

Engineers rearrange "the materials and forces of nature" to improve the way things work (Petroski, 1996, p. 1). Engineers solve problems like these: how to erect a suspension bridge that withstands high winds, how to design a lighter airplane or a smaller pacemaker, how to boost rocket thrust on a space shuttle. The engineer's concern is usually with practical applications, with structures and materials that are tested for safety and dependability.

Executives focus on decision making. In a global business climate of overnight developments (world markets, political strife, military conflicts, natural disasters), executives must often react on the spur of the moment. In such cases they rely on the best information immediately available—even when this information is incomplete or unverified (Seglin, 1998, p. 54).

Managers oversee the day-to-day operations of their organizations, focusing on problems like these: how to motivate employees, how to increase productivity, how to save money, how to avoid workplace accidents. They collaborate with colleagues and supervise various projects. To keep things running smoothly, managers rely on memos, reports, and other forms of information sharing.

Lawyers focus on protecting the organization from liability or corporate sabotage by answering questions like these: Do these instructions contain adequate warnings and cautions? Is there anything about this product that could generate a lawsuit? Have any of our trade secrets been revealed? Lawyers carefully review documents before approving their distribution outside the company.

The public focuses on the big picture—on what pertains to them directly: What does this mean to me? How can I use this product safely and effectively. Why should I even read this? They rely on information for some immediate practical purpose: to complete a task (What do I do next?), to learn more about something (What are the facts and what do they mean?), to make a judgment (Is this good enough?).

Because audiences' basic concerns vary, every audience expects a message tailored for its own specific interests and information needs.

 REVIEW CHECKLIST

Category	Questions
Analyzing audience	Who are they and what do they want to know?
Analyzing purpose	Why is this information needed? What tasks will be performed with it?
Analyzing context	What organizational, social, legal, or professional issues apply?
Other sources	
Corporate style guides	Is there a style guide available for reference?
User–preference documents	Do any user–preference documents exist?
Marketing surveys	Is any audience information available from the marketing department?

EXERCISES

1. Select a topic with which you are familiar; choose from hobbies, your job, or your academic major. Assume that you will be writing a brochure on this topic and that your audience is your classmates. Using the audience analysis worksheet, interview two or three classmates. From your notes, write an audience and purpose statement. It should begin: The audience for my brochure is [describe them]. The purpose of my brochure is [describe in terms of verbs: to inform, to train, to convince, etc.]. Trade your statement with a classmate and exchange feedback.

2. FOCUS ON ⮞WRITING. Based on your experience with Exercise 1, modify the audience analysis worksheet to include other questions, categories, or topics that you would need to learn about in order to understand your audience more fully. Write a memo to your instructor about your findings.

3. FOCUS ON ⮞WRITING. Locate a short article related to your major (or part of a long article or a selection from your textbook for an advanced course). Choose a piece written at the highest level of technicality you can understand. Using the audience analysis worksheet, write down the assumptions about the audience made by the author of this piece. What kind of audience did the author have in mind? What audience characteristics did she or he assume? What is/are the purpose(s) of this document? Now, working with a partner in class, discuss a different audience (lay people, mixed audience, novices) for this topic. Write about the changes you might make to turn this article into something accessible to a new audience.

4. Locate an example of technical communication (brochure, Web page, report) that you feel has both primary and secondary audiences, *or* locate an example that you feel has several purposes. Explain these audiences and purposes to your class. Is the material designed primarily for one audience or one purpose? How are the secondary audiences or purposes addressed?

The Collaboration Window

Form teams of 3–6 people. If possible, teammates should be of the same or similar majors (electrical engineering, biology, graphic design, etc.). Address the following situation: An increasing number of first-year students are dropping out of the major because of low grades, stress, or inability to keep up with the work. Your task is to prepare an online "Survival Guide" for incoming students. The Web site should focus on the challenges and pitfalls of the first year in this major. But before you can prepare the guide, you need to do a thorough analysis of its audience and purpose.

Assuming that some of you are in this major, perform an audience analysis using the worksheet on page 23 or a modified version of this worksheet. One team member should take notes, but all team members should participate, alternately, both as interviewers and interviewees. Take turns interviewing each student one at a time. Once you have a reasonable amount of information, draft an audience and purpose statement for your online Survival Guide (see Exercise 1).

The Global Window

Locate a document in hard copy or electronic form that appears to be designed primarily for U.S. audiences (it is written in English only, for example). Make a list of any features in this document that might reflect the writer's or designer's cultural bias. For example, items that U.S. readers may take for granted, such as the way dates are written, units of measurement, idioms, or slang, may be important issues if the document is read by someone who speaks American English as their second language. Discuss how you could edit these features so they would be more accessible to an international audience.

Click on This

Every day, people from all walks of life and backgrounds rely on weather reports. Meteorology is a scientific subject, yet most of us seem to understand such technical concepts as "high pressure system," "jet stream," and "cold front." Examine three or four weather-related Web sites and determine how the writers and designers assessed their audience and what devices they used to make this information understandable to nontechnical audiences. You can start with the Franklin Institute Science Museum (www.fi.edu/weather). Make a list of features that appeal to a mixed audience, and present your findings to class.

CHAPTER 3

Designing Usable Information

Usability and technical information

Usability—during the planning stages

Usability—during the writing and design process

Usability—after the information is released

Writing and organizing information for usability

Review checklist

Exercises

The collaboration window

The global window

Click on this

Usability and technical information

According to one report, companies "are finding that they must do more 'real-life' user testing to make their products useful to the average consumer" (*Investor's Business Daily,* 1998). When you plan, write, and design a piece of communication—a brochure, manual, online help screen, or report—you are creating a communication product. Like any other product, people will use it only if they can find what they need, understand the language, follow the instructions, and read the graphics. In other words, communication products must be *usable.* According to one group of experts, usability means that "people who use the product can do so quickly and easily to accomplish their own tasks" (Dumas & Redish, 1994, p. 4).

To create a usable communication product, begin with a careful audience and purpose analysis (Chapter 2). This analysis will provide you with much of the basic information you need to design your material. But you can take other, more specific steps to ensure that your document is usable during the planning stages, during the writing and design process, and after the release of your document.

Usability—during the planning stages

Before you begin writing or designing any information product, learn all you can about your audience and their intended use of your document. Then, develop a clear plan.

Perform an audience and purpose analysis

A systematic audience analysis is critical to any successful technical communication. To perform an audience analysis, customize the worksheet provided in Chapter 2 to fit your specific situation.

Perform a task analysis

Most audiences approach technical communication material with a series of tasks in mind. These tasks are most evident when the document is a set of instructions: Users want to install a new oil filter, assemble a new gas grill, or install a new word-processing program. But other forms of communication, such as reports, memos, and brochures, also involve user tasks. When reading a report, a manager may need to extract information and write a response. When replying to a memo, a technician may need to make an argument as to why the company should purchase new equipment. In this way, most technical communication is task-oriented. People come to the information wanting to *do* something.

As one team of experts notes, "[I]t is all too easy to forget that the product exists because human beings are trying to accomplish tasks. Task analysis refocuses attention on users, and on their tasks and goals" (Dumas & Redish, 1994, p. 44).

For your task analysis, you can create a worksheet similar to the one shown in Figure 3.1. Begin by defining the main tasks. For example, for an instruction manual to accompany a gas grill, you might define the primary user task as "assemble the grill." But this task can be divided into several smaller tasks. Note that the tasks are listed using verb forms (assemble, locate, get, and so on).

Assemble the grill:

1. Locate all parts.
2. Get the required tools.
3. Lay out parts in order.
4. Assemble parts into smaller units.
5. Assemble these smaller units into large unit.

Even this list can be subdivided: "assemble parts into smaller units," for example, probably consists of several smaller steps.

Main task: Assemble the grill
Subtasks
1. Locate all parts.
2. Get the required tools.
3. Lay out parts in order.
4. Assemble parts into smaller units.
5. Assemble these smaller units into large units.

Main task: Use the grill
Subtasks
1. Attach the gas canister.
2. Turn on the main gas valve.
3. Turn on the individual burners.
4. Press to ignite button.

Main task: Maintain the grill
Subtasks
1. Turn off the main gas valve when not in use.
2. Cover to protect from rain.
3. Clean the grate regularly.

Figure 3.1 ■ Sample task analysis worksheet for a gas grill instruction manual.

You can determine these tasks by interviewing users and watching them perform the actions. Once you understand the tasks to be performed, you can create an information plan (discussed next). Ultimately, your document will be more useful if you know what your audience wants to *do* with it.

Develop an information plan

Once you have a clear picture of the audience and purpose for your document, as well as the intended user tasks, you can draft an information plan: an outline of the assumptions, goals, specifications, and budget for your document. Information plans can be as short as a 2–3 page memo or as long as a 5–10 page report, depending on your project. Begin with a clearly stated goal ("users will be able to assemble a gas grill within 30 minutes") so that you can measure when a task has been successfully completed (Rubin, 1994, p. 97). Figure 3.2 is a sample information plan created in a short memo format.

Do the research

Developing an information plan might require research. For instance, you might need to determine how often gas grill accidents occur because the burner unit was assembled incorrectly or because a faulty connection has been overlooked. Such data will certainly affect your decisions about what information to include and what to leave out. For more information on performing research, see Chapter 4.

Usability—duRiNq thE wRiTiNq ANd dEsiqN pRocEss

Once you have completed these first steps, you can write, design, and test your document. For a gas grill instruction manual, you would write the instructions, design the graphics, and select a medium (print, CD-ROM, Web) for distributing the information. Most instructions are printed on paper and included with the product. You might choose paper but also make the information available on a Web site. Beside choosing a delivery method, you can take other steps at this stage to ensure usability.

Test early versions of your communication product

Allow audience members to provide input as early as possible. As one usability professional puts it, "[W]hen you involve your users as design partners, you create a partnership of ongoing usability testing throughout the information cycle" (Coe, 1996, p. 192). If time and budget allow, test your first draft of the

GrillChef Corporation

To: Technical writing design team
From: Erin Green and Geoff Brannigan, team leaders
Date: January 21, 2000
Re: Information plan for gas grill manual

As you know, our team recently performed an analysis of user needs as we prepare to design and write the new User Manual for the new GrillChef Model 2000 double-burner grill. This memo summarizes our findings and presents a plan for proceeding.

Part One: Analysis

Audience—The audience for this manual is very broad. It consists of consumers who purchase the grill. This purchase may be their first gas grill, or they may be replacing an old grill. Some users are making a switch from charcoal to gas. Our analysis revealed that the primary users are male and female, ranging in age from 25–50. From a focus group, we determined that most users are afraid to assemble the grill. But all members expressed enthusiasm about using the grill. Also, according to marketing, this grill is only sold in the United States.

Purpose—The manual has several purposes:
1. Instruct the user in assembling and using the grill.
2. Provide adequate safety instructions. These are to protect the user and to make sure we have complied with our legal requirements.
3. Provide a phone number, Web address, and other contact information if users have questions or need replacement parts.

User tasks—Our task analysis revealed three main tasks this manual must address:
1. How to assemble the grill. Users need clear instructions, a list of parts, and diagrams that can assist them. Users wish to be able to assemble the grill within 30 minutes to one hour.
2. How to use the grill. Users need clear instructions for operating the grill safely. Because some users have never used gas for grilling, we need to stress safety.
3. How to maintain the grill. Users need to know how to keep the grill clean, dry, and operational.

Part Two: Design Plans

Based on our analysis, we suggest designing a manual that is simple, easy to use, and contains information users need. We will follow the layout and format of our other manuals.

Rough outline—Cover with drawing of grill, model number, company name.
Inside front cover: safety warnings (our legal department has indicated that these warnings need to go first).
First section: Exploded diagram, list of parts, drawings of parts, numbered list of instructions for assembly.
Second section: Numbered list of steps for using the grill, accompanied by diagrams.
Third section: Bulleted list of tasks users must perform to maintain the grill.
Final page: Company address, phone number, and Web address.

Production guidelines—Our budget for this project will not allow for color printing or any photographs. We suggest black ink on white paper, 8-1/2 x 11 folded in half vertically. We can use line drawings of the Model 1999 and modify these to the specifications of the Model 2000.

Schedule—The manual must be ready for shipping on April 1, 1999. We will follow our usual production and writing schedule, briefly summarized here:
February 21: First draft of manual is complete. Manual is usability tested on sample customers.
March 1: Manual is revised based on results of usability test.
March 3: Copyediting, proofreading, and final changes. Manual goes to the printer.
March 30: Manual is back from printer and sent to the warehouse.
April 1: Product is shipped.

Figure 3.2 ■ A sample information plan for the gas grill manual.

brochure, Web page, or report on potential users. Ask people to read the material and "think out loud" about what they find useful and what they find confusing. Watch people use the material and see where your information works and where it fails. If someone were trying to assemble a gas grill but could not locate a part because of unclear instructions, this information would be invaluable as you revise the material. Ask for suggestions about graphics you might use, the format of your document, or the level of technical information.

Revise your plan and your product

Revise your information plan and your draft documents to conform to audience feedback. If your audience finds a technical term hard to understand, define it clearly or use a simpler word or concept. If a graphic makes no sense, locate one that does. It is easier to make these changes earlier rather than later, after your information has already been printed, distributed, or posted to the Web.

Create layered communication

A useful way to reach mixed audiences is to create "layered communication." If you have ever looked at the documentation that accompanies any word-processing software, you have seen layered communication. Since word-processing users range from novices to experts, one single type of documentation simply won't suffice. Novices would be confused by the shorthand and technical terms that make sense to experts, while experts would be frustrated with the level of detail and explanation needed for a novice user.

Word-processing documentation is often designed as a series of materials: a quick reference card, a "getting started" or "quick start" short manual, and a full-blown large manual. Depending on their level of expertise, users can scan these different materials, choosing the "layer" that matches their needs. Figure 3.3 shows an example of layered documentation for a product called "Conflict Catcher." The User Guide is primarily for novice users who want to learn about the product. The Quick Reference Guide is for users who know the product but need a quick reminder. The Troubleshooting Guide is for both experienced and novice users who don't need the full reference material but need a quick source of answers to solve specific problems.

Usability—After the information is released

Even after your instruction manual is on its way to the new gas grill owners or your report is being circulated among other engineers and managers, there are still ways to ensure usability in your information.

Figure 3.3 ■ Layered documentation.
Source: Courtesy Casady & Greene.

Provide mechanisms for user feedback

You can include ways for users to provide feedback on the documentation: customer comment cards, email addresses, phone numbers, Web sites. If your instructions contain a mistake on page six, you can be sure customers will let you know, provided you give them a way to reach you.

Plan for the next version or release

Continue collecting information, researching, and gathering user input as you plan the next version or release of your document. If you will need to write a revised manual in several months, begin collecting data as soon as possible. If the gas grill will be redesigned for next season, learn about the new design so you can plan the new manual. Establish an information file for quick access when you begin revising and updating.

Writing and organizing information for usability

You can dramatically increase the usability of any communication by focusing on three aspects of writing. First, use good grammar and style. Readers can't extract what they need from poorly written information. Moreover, bad writing makes you (and your company) look incompetent. Second, create an overview to give your audience a framework for navigating the document. Third, "chunk" your information into units that make sense for the specific audience and purpose. (Chunking is described later in this chapter.)

Using appropriate grammar and style

Following is a snapshot of important grammar and style issues for technical writing. For more information, see Appendix A for grammar issues and this book's companion Web site.

Use proper punctuation. A poorly punctuated sentence can be hard to interpret. One example is the use of the "series comma," which is a comma inserted before a coordinating conjunction in a list of items. For example:

> The jaw orb-weaver, wood spider, and lynx spider are examples of biological diversity in spiders.

Inserting a comma in the phrase "wood spider, and lynx spider" indicates that there are three items (jaw orb-weaver, wood spider, and lynx spider) in this series. But some writers, particularly in journalism, drop the final comma before the "and," as in:

> The jaw orb-weaver, wood spider and lynx spider are examples of biological diversity in spiders.

This usage suggests that "wood spider and lynx spider" are one unit, which is not the case.

Punctuation is easy once you learn some basic rules. Refer to the chart in Appendix A as you work on your own writing.

Use active voice often. In general, readers learn more quickly when communications are written in active voice. In active voice sentences, a clear agent performs a clear action on a recipient, as in:

Active voice	*Agent*	*Action*	*Recipient*
	(subject)	(verb)	(object)
	Joe	lost	your report.

Passive voice, on the other hand, reverses this pattern, placing the recipient of the action in the subject slot:

Passive voice	*Recipient*	*Action*	*Agent*
	(subject)	(verb)	(prepositional phrase)
	Your report	was lost	by Joe.

Note that passive voice adds a form of the verb "to be" (*was*) next to the actual verb.

Some writers mistakenly rely on passive voice because they think it sounds more objective and important. But passive voice decreases usability by making sentences wordier and harder to understand.

Writers often use passive voice to obscure the agent by leaving out the final phrase, as in:

Passive voice	*Recipient*	*Action*
	(subject)	(verb)
	Your report	was lost.

"Your report was lost" leaves out the responsible party. Who lost the report? Passive voice is unethical if it obscures the person or other agent who performed the action when that responsible person or agent should be identified.

Passive voice is appropriate when the agent is not known, or in cases where the object is more important than the subject. For example, if a group of scientists performed an experiment and wanted to explain the results, they might write

> The data were analyzed, and the findings were discussed.

Even here, the active voice ("We analyzed the data . . .") would be preferable. But if it was clear who analyzed the data, or truly not important to know who did the work, passive voice might be acceptable. Passive voice can also be used to ease the blow that a direct sentence might deliver; for example:

❚ You have not paid your bill.

The passive form is indirect and thus less offensive:

❚ Your bill has not been paid.

Consider this technique when you want to avoid a hostile tone. But in general, to create usable, readable technical information, use active voice.

Avoid nominalizations. A nominalization is a noun that would be easier to understand as a verb. Verbs are generally easier to read because they signal action. You can usually spot a nominalization in two ways.

  1. Look for words with a *-tion* ending:

    ❚ My recommendation is for a larger budget.

    Strike the *-tion,* to find the root verb: recommend. Then rewrite the sentence in a more direct form:

    ❚ I recommend a larger budget.

    Nominalizations may sound more "important" than a simpler verb form. But this kind of abstraction makes for difficult reading. A usable document is a readable document.

  2. Beware "the _____ of _____" formula:

    ❚ The managing of this project is up to me.

    This sentence is wordy and cumbersome. Locate the root verb form ("manage") and create a more accessible sentence, such as:

    ❚ I manage this project.

    or

    ❚ Managing this project is my job.

Unpack nouns. Too many nouns in a row can also create confusion and reading difficulty. One noun can modify another (as in "software development"). But when two or more nouns modify a noun, the string of words becomes hard to read and ambiguous. For example:

❚ Be sure to leave enough time for today's training session participant evaluation.

Is the evaluation of the session or of the participants? With no articles, prepositions, or verbs, readers cannot sort out the relationships among the nouns. Revise these sentences for clarity and readability:

▐ Be sure to leave enough time for participants to evaluate today's training session.

or

▐ Be sure to leave enough time to evaluate the participants in today's training session.

Avoid wordy phrases but don't overedit. Wordy phrases can often be reduced to one word:

at a rapid rate	=	rapidly
due to the fact that	=	because
aware of the fact that	=	know
in close proximity to	=	near

But don't overedit, leaving out so many words that your audience cannot follow your line of thinking. A sentence such as

▐ Proposal to employ retirees almost dead.

is confusing. What or who is "almost dead": the proposal or the retirees? A few more words would help:

▐ The proposal to employ retirees is almost dead.

Short sentences are good, but not at the expense of clarity. Clear information is usable information.

Use parallel structure. Parallel structure is a fancy way of saying that similar items should be expressed in similar grammatical form. For example, the following sentence is not parallel:

▐ She enjoys many outdoor activities, including running, kayaking, and the design of new hiking trails.

This sentence is essentially a list of items. The first two items, running and kayaking, are expressed as verbs with *-ing* endings. The third item, the design of new hiking trails, is not a verb, but a nominalization. To make this sentence parallel, you would revise as follows:

▐ She enjoys many outdoor activities, including running, kayaking, and designing new hiking trails.

Avoid useless jargon. Every profession has its own shorthand and accepted phrases and terms. Among specialists, these terms are an economical way to communicate. For example, "stat" (from the Latin "statim" or "immediately") is medical jargon for "drop everything and deal with this emergency." For computer engineers, a "virus" is not the common cold but a program that makes its way onto a computer's hard drive and causes problems.

Jargon can be useful when you are communicating with specialists. But some jargon is useless in any context. For example, the simple sentence

▮   We will cooperate on this project.

becomes jargon-filled in this next example:

▮   We will contiguously optimize our efforts on this project.

Only use jargon that improves your communication, not useless jargon that bogs down the information and sounds pretentious. Keep in mind that general audiences are unlikely to know the meaning of jargon that experts use. Depending on the situation, you will need to explain such terms or avoid using them altogether.

Avoid biased language. Language that is offensive or makes unwarranted assumptions will put off readers and make your document less effective. Women, for example, who receive a letter addressed to "Dear Sir" will probably throw the letter out and never read the information. Avoid sexist usage such as referring to doctors, lawyers, and other professionals as "him" or "he" while referring to nurses, secretaries, and homemakers as "her" or "she." Words such as *mailman* or *fireman* automatically exclude women, but words such as *mail carrier* or *firefighter* are far more inclusive.

Also, usable communication should respect all people regardless of their specific cultural, racial, or national backgrounds; sexual and religious orientations; ages or physical conditions. References to individuals and groups should be as neutral as possible. Avoid any expression that is condescending or judgmental or that violates the reader's sense of appropriateness.

Write from the user's point of view. As one team of technical communicators notes, "Writing from the user's point of view brings the user into the 'story,' so it is easy for the user to imagine doing what you are describing" (Hargis, 1997, p. 14). Techniques include performing an audience and purpose analysis (so you understand where your audience is coming from), writing in active voice, and creating headings in the form of reader questions. Another technique noted by Hargis is to use "you" (second person) or "you" understood whenever possible. Sentences such as "Insert the bolt into the large wheel frame" speak directly to the reader, not to some abstraction, such as "The bolt is now inserted. . . ."

Don't rely solely on grammar and spelling checkers. Some people mistakenly assume that the computer can solve all grammar and spelling problems. This is simply not true. Both "it's" and "its" are spelled correctly, but only one of them means "it is." The same is true for "their" and "there" ("their" is the possessive form, as in "their books," while "there" is an adverb, as in "There is my mother"). Spelling checkers are very important, because they will find words that are spelled incorrectly, but don't count on them to find words that are *used* incorrectly. Grammar checkers are also fine tools to help you locate possible problems, but do not rely on what the software tells you. For example, not every sentence that the grammar checker flags as "long" should be shortened. Use these tools wisely and with common sense. Also, ask someone to proofread your material. Some companies have full-time technical editors who are happy to look over your writing.

Consider international issues and writing for translation. Technical communication is a global process. Documents may originate in English but then be translated into other languages. In this case, writers must be careful to use English that is easy to translate. Idioms, humor, and analogies are often difficult for translators. One famous example is the case of a U.S. car called the Nova. When translated into Spanish, Nova means "Does Not Go"! In addition to terms, certain grammatical elements are also important for translation. The lack of an article (*a, the*) or of the word *that* in certain crucial places can cause a sentence to be translated inaccurately. Consider the following example (Kohl, 1999):

> Programs **that are** currently running in the system are indicated by icons in the lower part of the screen.

> Programs currently running in the system are indicated by icons in the lower part of the screen.

The first sentence contains the phrase "that are," a phrase commonly left out by native English writers, as in the second sentence. This second sentence may make sense, but it is hard to translate, because the phrase "that are" provides the translator with cues about the meaning of "running."

Use white space and effective page design. The proper use of white space, typography, and page design plays a big role in a document's usability. Pages that are set in tiny type with text crammed on the page and no graphics are hard to read. Chapter 8 provides more information on graphics and visuals.

Creating an overview

Information is usable when people can answer several key questions:

- What will I learn from this document?
- Why am I receiving this information?
- What can I anticipate finding in this document?

To help answer these questions, always provide an overview before launching into the details. Think about it this way: If you were taking a long road trip, you would probably study a map first to get the "big picture" of your journey and to know exactly where you will be headed. Overviews provide a sort of road map.

You can provide an overview in many forms and places within a document. Some documents begin with a section called "About This Document," which previews the entire document. Within a document, you can also provide an overview of each chapter. Figure 3.4 shows a book overview from a manual for an IBM laptop computer. This particular overview explains what the audience will learn, how long this process should take, and what steps they should already have completed.

Overviews are important in oral communication as well. At the beginning of a presentation, preview for your audience the main points you will be covering. For example, you might begin a presentation about electric cars by saying, "Today, I would like to give you more information about electric cars. Specifically, I will cover three main points: the way electric cars operate, certain new designs in electric batteries, and the usefulness of electric cars in cold climates." Your presentation would then cover these points, in that order.

Chunking information

"Chunking" is an organizing technique in which you divide the information into small units or modules based on the topics or types of information that will be covered in a given section. When you chunk information into topics, you should "include information in a topic that the user thinks of as a unit" (Horton, 1990, p. 101).

If you were designing a quick reference card for using an ATM, you might discover that the information falls into three general topics, or chunks:

- How to make a withdrawal
- How to make a deposit
- How to check your balance

You would design the card around these three chunks of information. If any one of the chunks became too long and unwieldy, you might subdivide it.

About this book

This book contains information that will help you operate the IBM ThinkPad 240 computer. Be sure to read the *ThinkPad 240 Setup Guide* and Chapter 1 of this book before using the computer.

Chapter 1. "Getting Familiar with Your Computer," acquaints you with the basic features of your computer.

Chapter 2. "Extending the Features of Your Computer," provides information on installing options and using your computer's high-technology features.

Chapter 3. "Protecting Your Computer," provides information on using passwords, and using locks.

Chapter 4. "Solving Computer Problems," describes what to do when you have a computer problem. The chapter includes a troubleshooting guide on how to recover lost or damaged software.

Chapter 5. "Getting Service," describes various options of IBM's support and service.

Appendix A, Features and specifications describes the features and specifications associated with your computer, including information on power cords.

Appendix B, Product warranties and notices contains the warranty statements for your computer and notices for this book.

The **glossary** defines terms appearing in this book. The book concludes with an index.

v

FIGURE 3.4 ■ **Book overview from the IBM Thinkpad 240 manual.** This page is designed to help users understand the purpose and structure of the manual.
Source: IBM Thinkpad 240 User's Reference, First Ed., 1999.

"How to make a deposit," for example, might become two chunks: "How to deposit checks" and "How to deposit cash."

If you've ever created an outline to help you write a paper or speech, you've had experience in breaking down information into smaller units. When you chunk information for an audience, you create these units based on the audience's needs and the document's purpose.

Creating headings

Another way to enhance usability is to create headings in the form of questions your audience might ask. This approach isn't appropriate for all documents, and overuse of questions can become repetitive and annoying. For certain documents, though, such as patient information brochures, questions can help guide readers to the appropriate section of the document. Questions also create an inviting, user-friendly tone: If the question sounds like something readers would actually ask, they will feel as if the document has been written just for them.

For example, a patient information brochure about a laparoscopy (a medical procedure that uses a small camera to look inside the body) begins as follows:

> **LAPAROSCOPY**
>
> **Activities**
>
> You should rest until you feel up to resuming your normal activities—usually in a day or two. Do not lift objects weighing more than 20 to 30 pounds for one week.
> *Source:* University of Minnesota Hospital and Clinics.

This information would be more useful and friendly if it addressed an actual patient question:

> **WHAT ACTIVITIES CAN I PERFORM AFTER MY LAPAROSCOPY?**
>
> You should rest until you feel up to resuming your normal activities—usually in a day or two. Do not lift objects weighing more than 20 to 30 pounds for one week.

Using the margins for commentary

You can use the margins to call out or highlight particularly important information, or you can leave them blank for readers to take notes or write comments (see Figure 3.5). Using white space and marginal cueing areas are discussed in more detail in Chapter 8.

Review Checklist

Achieving usable communication	Strategies
Planning for usable communication	Perform an audience and purpose analysis.
	Perform a task analysis.
	Develop an information plan.
	Do the research.
Writing and designing usable communication	Test early versions of your communication product.
	Revise your plan and your product.
	Create layered communication.
Writing and organizing your information with usability in mind	Use appropriate grammar and style.
	Create overviews.
	Chunk the information.
	Create headings in the form of audience questions.
	Use margins for commentary.
	Use white space and effective page design.
Following up on your communication	Provide mechanisms for user feedback.
	Plan for the next version or release.

Exercises

1. Identify an activity that could require instructions for a novice to complete. Prepare a task analysis for this activity using a worksheet similar to the one in Figure 3.1. Exchange task analyses with another student in your class and critique one another's analyses. With your class, discuss the challenges of doing a task analysis and identify strategies for performing a task analysis effectively.

2. Focus on ⟶Writing. Find a set of instructions or another type of technical document that is easy to use. Identify specific characteristics of the document that make it usable. Is the document well written? Does it have an overview? Can you quickly find the information you need? Then, find a technical document that is hard to use. What characteristics make it unusable? In a memo to your instructor, define specific changes that you would make in revising the document. Submit both examples along with your memo.

3. Focus on ⟶Writing. The sentences on page 48 violate many of the grammar and style guidelines that you studied in this chapter. Revise each sentence to improve its usability.

Increasing memory

Increasing memory capacity is an effective way to make programs run faster. Your dealer can increase the amount of memory in your computer by installing a small outline dual inline memory module (SoDIMM), available as an option.

Different capacities of SoDIMM are available.

Notes:

1. If you changed the memory installed in the computer, you need to create a new hibernation file. To create the hibernation file

   ➡ the Online User's Guide.

Upgrading the memory

You may upgrade the memory of the ThinkPad 240. Please refer to the following illustration and instructions:

1. Turn the power off. Disconnect all peripherals. Disconnect the AC adapter.

2. Remove the battery (see "Replacing the battery pack" on page 34).

3. Open the LCD panel to an angle of between 90 to 130 degrees, and set the computer on its side.

4. Locate the three screw holes that are marked with a memory chip icon ■■. Holding the keyboard with your hand as shown in the illustration below, remove the three screws from these holes and set them aside.

Figure 3.5 ■ **Page from a manual for an IBM ThinkPad 240.** Note how the page is designed with room in the left margin for users to make notes.
Source: IBM Thinkpad 240 User's Reference, First Ed., 1999.

a. Your correspondence has been duly forwarded for consideration by the personnel office, which has employment candidate selection responsibility.

b. Uninsulated end pipe ruptured.

c. Do not enter test area while contaminated.

d. Develop online editing system documentation.

e. Sarah's job involves fault analysis systems troubleshooting handbook preparation.

f. Labor costs for this job were underestimated.

g. A layoff is recommended.

h. If my claim is not settled by May 15, the Better Business Bureau will be contacted and their advice or legal action will be taken.

i. Care should be taken with dynamite.

j. It was decided to reject your offer.

 THE COLLABORATION WINDOW

Bring in some children's connecting blocks, such as TinkerToys or Lego blocks. Form several teams of four to six people, and assign two people as technical writers. The two people assigned as technical writers should assemble a few of the pieces into a simple design (don't use more than three or four pieces). Now, proceed as follows:

The technical writers should write up a quick instruction card explaining how to assemble the pieces into the design they've created. For example, the card for Lego blocks might say:

1. Locate two large red blocks and two small green blocks

2. Place one red block on its side

3. Attach one green block to the red block

and so on. Now, present the "parts" and your instructions to the rest of your team. Watch as your team tries to assemble the blocks according to your instructions. Assess the instructions for usability. Were all tasks accounted for? Did any terms or language confuse the users? Go back and perform a task analysis, and discuss what you could do to improve the usability of your instruction card.

The Global Window

Focus on ▪▪▪➡ Writing. If you have ever seen the image of the earth rising above the moon, photographed by one of the astronauts on a moon landing, you have probably experienced the feeling that the world is truly one home for all of us. Regardless of our language, culture, or physical location, we all reside on this planet. Knowledge about astronomy and planetary science belong to everyone, not just one country. Yet many of the communications about space are written in English and are accessible only to those who understand this language. Using a search engine, locate the Web site for the Planetary Society and other space-related organizations. Explore several of these sites. Assess the usability for international audiences of three of these Web sites. Would non-English speakers be able to access the information contained on these pages? Present your evaluation in the form of a memo to your instructor, accompanied by a printout of each of the three home pages.

Click on This

Usable communication products must use terminology that is consistent with the audience's background and level of understanding. Often, as a technical communicator, you will encounter technical terms that need to be revised for a nonexpert audience. But what can you do if you do not understand the technical term itself? There are several good online dictionaries you can turn to for help. Try the Web site dictionary.langenberg.com. Look in a technical or science journal and find a term that you do not understand, and see if you can locate its meaning in this dictionary.

Another useful Web site is the Usability Professionals Association located at www.upassoc.org.

CHAPTER **4**

Performing Research for Technical Communication

Thinking critically about research

Primary research

Internet research

Other electronic research tools

Hard copy research

Review checklist

Exercises

The collaboration window

The global window

Click on this

Thinking critically about research

Most major decisions in technical communication are based on careful research, often with the findings recorded in a written report, in a long memo, on a Web site, or in some combination of documents. The type of research you will perform as a technical communicator depends largely on your workplace or classroom assignment. For any topic you research, for the classroom or workplace, you can consult numerous information sources. An excellent place to begin is with primary research, where you can get information "from the source" by conducting interviews and surveys and by observing people in action. The Internet is also a great choice, because it's quick and convenient, but you need to verify that the information you find on the Internet is from a credible source. Other electronic sources include CD-ROM databases and online retrieval services. Finally, traditional sources such as encyclopedias, print indexes, and journals can be valuable because their contents are usually subject to close scrutiny before they are published. Many of these traditional sources are also available electronically.

Primary research

Informative interviews, surveys, questionnaires, inquiry letters, observations/experiments, and official records are considered *primary sources* because they afford an original, firsthand study of a topic.

Informative interviews

An excellent primary source of information is the interview, conducted either in person, by telephone, or by email. Much of what an expert knows may never be published. Also, a respondent might refer you to other experts or sources of information.

Of course, an expert's opinion can be just as mistaken or biased as anyone else's. Like patients who seek second opinions about serious medical conditions, researchers seek a balanced range of expert opinions about a complex problem or controversial issue—not only from a company engineer and environmentalist, for example, but also from independent and presumably more objective third parties such as a professor or journalist who has studied the issue.

Always go into an interview with a clear purpose and do your homework before the interview so you won't waste time asking questions you could have answered yourself.

INTERVIEW TIPS

- Make each question clear and specific.
- Avoid questions that can be answered with "yes" or "no."
- Avoid loaded questions such as, "Wouldn't you agree that hazards about genetically modified foods have been overstated?" Ask impartial questions instead, such as "In your opinion, have genetically modified food hazards been accurately stated, overstated, or understated?"
- Save the most difficult, complex, or sensitive questions for last.
- Be polite and professional.
- Let your respondent do most of the talking.
- Ask for clarification if needed, but do not put words in the respondent's mouth. Questions such as "Could you go over that again?" or "What did you mean by *X*?" are fine.

Surveys and questionnaires

Surveys help you to develop profiles and estimates about the concerns, preferences, attitudes, beliefs, or perceptions of a large, identifiable group (a *target population*) by studying representatives of that group (a *sample*).

The questionnaire is the tool for conducting surveys. While interviews allow for greater clarity and depth, questionnaires offer an inexpensive way to survey a large group. Respondents can answer privately and anonymously—and often more candidly than in an interview. Follow the guidelines below to design effective surveys.

SURVEY AND QUESTIONNAIRE TIPS

- **Define the survey's purpose and target population.** Why is this survey being performed? What, exactly, is it measuring? How much background research do you need? How will the survey findings be used? Who is the exact population being studied (the chronically unemployed, part-time students, computer users)?
- **Identify the sample group.** How will intended respondents be selected? How many respondents will there be? Generally, the larger the sample surveyed, the more dependable the results (assuming a well-chosen and representative sample). Will the sample be randomly chosen? In the statistical sense, "random" does not mean " haphazard": a random sample means that each member of the target population stands an equal chance of being in the sample group.
- **Define the survey method.** What type of data (opinions, ideas, facts, figures) will you collect? Is timing important? How will the survey be administered—in person, by mail, by phone? How will the data be collected, recorded, analyzed, and reported (Lavin, 1992, p. 277)?

Phone, email, and in-person surveys yield fast results and high response rates, but respondents consider phone surveys annoying and, without anonymity, people tend to be less candid. Mail surveys are less expensive than phone surveys over long distances. Electronic surveys, conducted via a Web form or an email message, are the least expensive way to conduct surveys. But these methods can have pitfalls. Computer connections can fail, and you have less control over how many times the same person completes the survey.

- **Decide on the types of questions.** Questions can take two forms: open-ended and closed-ended. Open-ended questions allow respondents to answer in any way they choose. It is more time-consuming to measure the data gathered, but such questions provide a rich source of information. An open-ended question would be worded like this:

  How much do you know about genetically modified food products?

  Closed-ended questions give respondents a limited number of choices and the data gathered are easier to measure. A closed-ended question would be set up like this:

  Are you concerned about genetically modified food products?

  Yes _____   No _____

- **Develop an engaging introduction and provide appropriate information.** Persuade respondents that the questionnaire relates to their concerns, that their answers matter, and that their anonymity is ensured:

  Your answers will help our company determine the public's views about genetically modified food products. All answers will be kept confidential. Thank you.

- **Make it brief, simple, and inviting.** Respondents don't mind giving up some time to help, but long questionnaires usually don't get many replies. Limit the number and types of questions to the most important topics.

Public records and organizational publications

The Freedom of Information Act and state public record laws grant public access to an array of government, corporate, and organizational documents. Obtaining these documents (from state or federal agencies) takes time (although more and more such documents are available on the Web), but in them you can find answers to questions like these (Blum, 1997, pp. 90–92):

- Which universities are being investigated by the USDA (Dept. of Agriculture) for mistreating laboratory animals?
- Are IRS auditors required to meet quotas?
- How often has the local nuclear power plant been cited for safety violations?

Organization records (reports, memos, Web pages, and so on) are also good primary sources. Most organizations publish pamphlets, brochures, annual reports, or Web sites for consumers, employees, investors, or voters. Of course, you need to be alert for bias in company literature. In evaluating the safety of a genetically modified food product, you would want the complete picture. Along with the company's literature, you would want studies and reports from government agencies and publications from health and environmental groups.

Personal observation and experiments

Observation should be your final step in primary research because you now know what to look for. Know how, where, and when to look, and jot down observations immediately. You might even take photos or make drawings.

Unlike general observations, experiments are controlled forms of observations designed to verify an assumption (e.g., the role of fish oil in preventing heart disease) or to test something untried (e.g., the relationship between background music and productivity). Each field has its own guidelines for experimental design, including the need for all experiments to be reviewed by Human Subjects Review Boards to ensure protection of test subjects. Depending on the situation, you may be able to learn about your subject matter by interviewing a scientist who is conducting an experiment or by reading published results.

Finally, workplace research can involve the analysis of samples, such as water, soil, or air for contamination and pollution; foods for nutritional content; plants for medicinal value. Investigators may analyze material samples to find the cause of airline accidents; engineers may analyze samples of steel, concrete, or other building materials to test for tensile strength or load-bearing capacity. Medical specialists may analyze tissue samples for disease. As a researcher, you may be able to access this information through interviews or published reports.

Internet research

The Internet can provide sources ranging from 8-year-olds who have their own Web sites to electronic journals and national newspapers. Remember that while the Internet is quick and easy to use, you need to be sure that the source of your information is credible and reliable. Several kinds of Internet sources, along with tips for checking their reliability, are described below.

Usenet news

Usenet is a worldwide system for online discussions via newsgroups, a type of electronic bulletin board on which users post and share information and discuss topics of common interest via email.

Newsgroups are either *moderated* or *unmoderated.* In a moderated group, all contributions are reviewed by a moderator who must approve the material before it can be posted. In an unmoderated group, all contributions are posted. In using newsgroups for your own research, keep in mind that material posted to moderated newsgroups is generally more reliable than that posted to unmoderated newsgroups because the moderating process attempts to filter out messages containing unsubstantiated claims and other kinds of inappropriate material (Munger, Anderson, Benjamin, Busiel, Parades-Holt, 2000, p. 48). Most newsgroups are unmoderated. Newsgroups are useful because they make vast amounts of information immediately available. However, when using either moderated or unmoderated newsgroups, it's always a good idea to read with a critical eye and to verify your information with at least one other source.

Also available are *newsfeed* newsgroups, which gather and post news items from wire services such as the Associated Press. Newsfeed newsgroups can be particularly useful in researching a topic because they contain a wealth of reliable information. Whereas newspapers only have room to print a portion of the information they receive from wire services such as the Associated Press, newsfeed newsgroups make all this information available online.

Usenet Tips

- To search for newsgroups on any topic, go to www.liszt.com/news.
- Most Web browsers support newsgroups. If you know the name of the newsgroup you want to visit, type *news:* followed by the newsgroup name in your browser location box.
- Because the information in newsgroups is always changing, make sure you immediately save any postings that you find useful: They might not be there when you go back.

Listservs

A listserv is, quite simply, a computer-operated mailing list. Like newsgroups, listservs are special-interest groups for email discussion and information sharing. In contrast to newsgroups, listserv discussions usually focus on specialized topics, with discussions usually among experts (say, cancer researchers), often before their findings or opinions appear in published form. Many listservs include a FAQ listing.

Listservs that post items such as newspaper articles and expert commentary on specialized topics are particularly useful for research purposes. However, listservs that consist primarily of discussion among nonexperts can also provide a useful introduction to an issue if you are in the early stages of researching a topic (Munger et al., 2000, p. 19).

Listserv access is available to subscribers who receive mailings automatically via email. Like a newsgroup, a listserv may be moderated or unmoderated, but subscribers/contributors are expected to observe "netiquette" (proper Internet etiquette) and to stick to the topic without digressions, "flaming" (attacking someone), or "spamming" (posting irrelevant messages).

LISTSERV TIPS

- To find listservs on your topic, go to www.liszt.com/ or tile.net/lists.
- Once you find a listserv that you think might be useful, follow the instructions provided at the Web sites listed above to subscribe to the listserv. These instructions will tell you to send an email message to a particular address. Because listserv mailing lists are maintained by software, not by people, it's very important to follow the exact directions given for subscribing and unsubscribing.

Electronic magazines (zines)

Zines offer information available only in electronic form. Despite the broad differences in quality among zines, this online medium offers certain benefits over hard copy magazines:

- Links to related information
- Immediate access to earlier magazine issues
- Interactive forums for discussions among readers, writers, and editors
- Rapid updating and error correction

Major news publications and television news programs also offer interactive editions online. Some of these include the following:

New York Times (www.nytimes.com)

Washington Post (www.washingtonpost.com)

Time Magazine (pathfinder.com/time)

MSNBC (msnbc.com)

CNN (cnn.com)

Email inquiries

Email is excellent for contacting knowledgeable people in any field. Email addresses are increasingly accessible via locator programs that search various local directories listed on the Internet (Steinberg, 1994, p. 27). Keep in mind, though, that unsolicited and indiscriminate email inquiries might offend the recipient.

The Web

The Web is a global network of databases, documents, images, and sounds. All types of information from anywhere on the Web can be accessed and explored through navigation programs such as *Netscape Navigator* or *Microsoft Internet Explorer* (known as browsers). Hypertext links among Web resources allow users to explore information along different paths by clicking on key words or icons.

Each Web site has its own main page, which serves as an introduction to the site and is linked to additional pages that individual users can explore according to their information needs. There are two ways to find Web sites relevant to your topic: *subject directories* and *search engines.*

Subject directories. Subject directories are maintained by people who visit large numbers of sites and then organize links to the ones they find useful by maintaining lists of topic headings and subheadings. The oldest and most popular subject directory index is maintained by *Yahoo!* (www.yahoo.com). Other subject directory indexes are *The Internet Public Library* (www.ipl.org), *Library of Congress* (lcWeb.loc.gov), *WWW Virtual Library* (www.vlib.org), and *The Argus Clearinghouse* (www.clearinghouse.net). You can use these subject directory indexes either by navigating through the lists of headings and subheadings or by doing keyword searches.

Search engines. Search engines are maintained by computers. Search engines will locate a lot more information on your topic than subject directories because the people who maintain subject directories filter out sites that they deem irrelevant or unworthy. However, the amount of information gathered using a search engine can be overwhelming, so you have to do a lot of filtering to find the high-quality information most relevant to your topic (Munger et al., 2000, p. 32). Popular search engines include *AltaVista* (www.altavista.com), *Lycos* (www.lycos.com), *Excite* (www.excite.com), *InfoSeek* (infoseek.go.com), and *Google* (www.google.com).

SEARCH TIPS

Most engines that search by keyword allow the use of Boolean operators (commands such as "AND," "OR," "NOT," and so on) to define relationships among various key words. The table below shows how these commands can expand or narrow a search by generating more or fewer hits.

If you enter these terms ...	The computer searches for ...
electromagnetism AND health	only entries that contain both terms
electromagnetism OR health	all entries that contain either term

(continued)

If you enter these terms . . .	The computer searches for . . .
electromagnetism NOT health	only entries that contain term 1 and do not contain term 2
electromag*	all entries that contain this root within other words

Boolean commands can also be combined, as in

> (electromagnetic OR radiation) AND (fields OR tumors)

The hits produced here would contain any of these combinations:

> electromagnetic fields, electromagnetic and tumors, radiation and fields, radiation and tumors

Using *truncation* (cropping a word to its root and adding an asterisk), as in *electromag**, would produce a broad array of hits, including these:

> electromagnet, electromagnetic energy, electromagnetic impulse, electromagnetic wave

Different search engines use Boolean operators in slightly different ways; many include additional options (such as NEAR, to search for entries that contain search terms within ten or twenty words of each other). Click on the HELP option of your particular search engine to see which strategies it supports.

Regardless of your topic, a Web search typically yields a large number of hits. You need to evaluate every source you intend to use. If you can't clearly identify the author and his or her credentials, you should assume that the source is not credible. Why? Because the Web is a *bottom-up* medium, allowing anyone with the technical resources to create a Web site. Fonts, color, and images can make any site appear credible. Information comes from a multitude of sources, without any form of gatekeeping.

On the other hand, journal articles, newspaper reports, and television or radio stories are considered *top-down* information. These materials are usually subject to editorial or peer review and fact-checking before being printed or delivered on the air. And even though all information should be judged critically, Web sites require particular attention.

For example, imagine you are researching the topic of genetically modified organisms (GMOs); in particular, you are looking into the pros and cons of using genetically modified corn, oats, and other grains for food. You use a search engine and search on the term "GMO." One of your hits is the Web site shown in Figure 4.1.

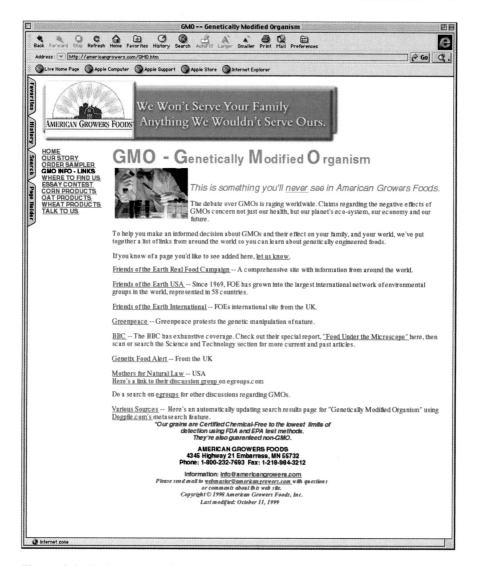

Figure 4.1 ■ **A Web site that advocates a particular viewpoint.** One of the Web site hits found during a search on the term "GMO."
Source: American Growers Foods Web site.

Too often, people seem to think that information located via a Web search can be cited without evaluating the source. This site has a clear source: a group called American Growers Foods. It also offers many links to information about GMOs. But does this site offer neutral information? The slogan "We won't serve your family anything we wouldn't serve ours" at the top is one clue that this site might be anti-GMO, which you can then confirm by noting the statement at the bottom ("Our grains . . . are guaranteed non GMO"). Do these statements mean the site is an invalid source of information? Not at all. They do mean, however, that this information source, like all others, advocates a point of view and should be considered as such. Balance your electronic research with sources such as academic journal articles, documentary radio or television programs, technical manuals, and reputable media reporting.

Walker and Ruszkiewicz (2000) suggest the following options for keeping track of the vast amount of information you are likely to find on the Internet:

- **Cut and paste.** You can copy URLs (World Wide Web addresses) and text from Web sites, newsgroups, or listservs directly into your word-processing or database files. This saves you time because you won't have to retype or rewrite information to integrate it into your own documents.
- **Download or print out files.** If you find information at a site that appears to change frequently or is hard to access, you should either download and save the information or print it out. Make sure you accurately record the URL where you found the information and the date on which you accessed the site. Most Web browsers can be configured to print this information when they print the page—you can usually specify this option in your browser's page setup menu.
- **Use your bookmarks file.** Most browsers provide some kind of bookmarking system that allows you to save and organize bookmarks to sites you've found useful. Typically, this feature allows you to create a system of folders to categorize the different kinds of sites that you locate. You can also save your bookmark files to floppy disks, which is especially useful if you work in a public computer lab.
- **Electronic note cards and bibliography programs.** Software programs such as *ProCite* or *Endnote* allow you to store complete bibliographic information as well as abstracts and quotations from various kinds of sources. This kind of software provides a useful system for managing all your research, not just Internet research.

Other electronic research tools

There are several other electronic technologies often used for storing and retrieving information. These technologies are accessible at libraries and, in some cases, via the Web.

Compact discs

A single CD-ROM disc can store the equivalent of an entire encyclopedia and serves as a portable database, usually searchable via key word. One useful

CD-ROM for business information is *ProQuest™*: Its *ABI/INFORM* database indexes over 800 journals in management, marketing, and business published since 1989; its *UMI* database indexes major U.S. newspapers. A useful CD-ROM for information about psychology, nursing, education, and social policy is *SilverPlatter™*.

Online retrieval services

College libraries and corporations subscribe to online services that can access thousands of databases stored on centralized computers. Compared with CDs, mainframe databases are usually more specialized and more current, often updated daily (as opposed to weekly, monthly, or quarterly updating of CD databases). Online retrieval services offer access to three types of databases: bibliographic, full-text, and factual (Lavin, 1992, p. 14):

- *Bibliographic databases* list publications in a particular field and sometimes include abstracts for each entry.
- *Full-text databases* display the entire article or document (usually excluding graphics) directly on the computer screen, and will print the article on command.
- *Factual databases* provide facts of all kinds: global and up-to-the-minute stock quotations, weather data, lists of new patents filed, and credit ratings of major companies, to name a few.

Four popular database services are:

- **OCLC and RLIN.** You can easily compile a comprehensive list of works on your topic at any library that belongs to an electronic consortium such as the Online Computer Library Center (OCLC) or the Research Libraries Information Network (RLIN). OCLC and RLIN databases are essentially giant electronic card catalogs. Using a networked terminal, you can search the databases by subject, title, or author.
- **DIALOG.** Many libraries subscribe to DIALOG, a network of more than 150 independent databases covering a range of subjects and searched by key words. This system can provide bibliographies and abstracts of the most recent journal articles on your topic. DIALOG databases include *Conference Papers Index, Electronic Yellow Pages (for Retailers, Services, Manufacturers),* and *ENVIROLINE*.
- **BRS.** Bibliographic Retrieval Services, another popular network consisting of more than 50 databases, provides bibliographies and abstracts from life sciences, physical sciences, business, or social sciences. BRS databases include *Dissertation Abstracts International, Harvard Business Review,* and *Pollution Abstracts*.

Comprehensive database networks such as DIALOG and BRS are accessible via the Internet for a fee. Specialized databases, such as *MEDLINE* or

ENVIROLINE, offer free bibliographies and abstracts, and copies of the full text can be ordered for a fee. Ask your librarian for help searching online databases.

Card catalog

Most library card catalogs aren't made up of cards anymore—rather, they are electronic and can be accessed through the Internet or at terminals in the library. You can search a library's holdings by subject, author, or title in that library's catalog system. Visit the library's Web site or ask a librarian for help.

Hard copy research

Traditional printed research tools are still invaluable. Unlike much of what you may find on the Web (especially if you aren't careful about checking the source), most print research tools are carefully reviewed and edited before they are published. True, it may take more time to go to the library and look through a printed book, but often, it's a better way to get solid information. In time, many of these sources will become available on the Web (many are now).

Bibliographies

These comprehensive lists of publications about a subject are generally issued yearly or even more frequently. However, they can quickly become dated. To see which bibliographies are published in your field, begin with the *Bibliographic Index,* which is a list (by subject) of bibliographies that contain at least 50 citations. To look for bibliographies on scientific and technical topics, consult *A Guide to U.S. Government Scientific and Technical Resources,* which is a list of everything published by the government in these broad fields. You might also look for bibliographies focused on a particular subject, such as *Health Hazards of Video Display Terminals: An Annotated Bibliography.*

Encyclopedias

Encyclopedias provide basic information. Examples include *Encyclopedia of Building and Construction Terms, Encyclopedia of Banking and Finance, Encyclopedia of Food Technology.* The *Encyclopedia of Associations* lists over 30,000 professional organizations worldwide (American Medical Association, Institute of Electrical and Electronics Engineers, and so on). Most organizations can be accessed via their Web sites.

Dictionaries

Dictionaries may be general or they may focus on specific disciplines or give biographical information. Examples include *Dictionary of Engineering and Technology, Dictionary of Telecommunications,* and *Dictionary of Scientific Biography.*

Handbooks

These research aids gather key facts (formulas, tables, advice, examples) about a field in condensed form. Examples include *Business Writer's Handbook, Civil Engineering Handbook,* and *The McGraw-Hill Computer Handbook.*

Almanacs

Almanacs contain factual and statistical data. Examples include *World Almanac and Book of Facts, Almanac for Computers,* and *Almanac of Business and Industrial Financial Ratios.*

Directories

Directories provide updated information about organizations, companies, people, products, services, or careers, often including addresses and phone numbers. Examples include *The Career Guide: Dun's Employment Opportunities Directory, Directory of American Firms Operating in Foreign Countries,* and *The Internet Directory.*

Guides to literature

If you simply don't know which books, journals, indexes, and reference works are available for your topic, consult a guide to literature. For a general list of books in various disciplines, see Walford's *Guide to Reference Material* or Sheehy's *Guide to Reference Books.* For scientific and technical literature, consult Malinowsky and Richardson's *Science and Engineering Literature: A Guide to Reference Sources.* Ask your librarian about literature guides for your discipline.

Indexes

Indexes are lists of books, newspaper articles, journal articles, or other works on a particular subject.

Book indexes. A book index lists works by author, title, or subject. Sample indexes include *Scientific and Technical Books and Serials in Print* (an annual listing of literature in science and technology), *New Technical Books: A Selective List with Descriptive Annotations* (issued ten times yearly), *Medical Books and Serials in Print* (an annual listing of works from medicine and psychology).

Periodical indexes. A periodical index provides sources from magazines and journals. First, decide whether you seek general or specialized information. Two general indexes are the *Magazine Index,* a subject index on microfilm, and the *Readers' Guide to Periodical Literature,* which is updated every few weeks.

For specialized information, consult indexes that list journal articles by discipline, such as *Ulrich's International Periodicals Directory,* the *General Science Index,* the *Applied Science and Technology Index,* or the *Business Periodicals Index.* Specific disciplines have their own indexes: Examples include *Agricultural Index, Index to Legal Periodicals,* and *International Nursing Index.*

Citation indexes. Citation indexes allow researchers to trace the development and refinement of a published idea. Using a citation index, you can track down the specific publications in which the original material has been cited, quoted, applied, critiqued, verified, or otherwise amplified (Garfield, 1973, p. 200). In short, you can use them to answer this question: *Who else has said what about this idea?*

The *Science Citation Index* cross-references articles on science and technology worldwide. Both the *Science Citation Index* and its counterpart, the *Social Science Citation Index,* are searchable by computer.

Technical report indexes. Government and private sector reports prepared worldwide offer specialized and current information. Examples include *Scientific and Technical Aerospace Reports, Government Reports Announcements and Index,* and *Monthly Catalog of United States Government Publications.* Proprietary or security restrictions limit public access to certain corporate or government documents.

Patent indexes. Countless patents are issued yearly to protect rights to new inventions, products, or processes. Information specialists Schenk and Webster (1984) point out that patents are often overlooked as sources of current information: "Since it is necessary that complete descriptions of the invention be included in patent applications, one can assume that almost everything that is new and original in technology can be found in patents" (p. 121). Examples include *Index of Patents Issued from the United States Patent and Trademark Office, NASA Patent Abstracts Bibliography,* and *World Patents Index.*

Patents in various technologies are searchable through databases such as *Hi Tech Patents, Data Communications,* and *World Patents Index.*

Abstracts

By indexing and summarizing each article, abstracts can save you from having to locate a journal before deciding whether to read the article or skip it. Abstracts are usually titled by discipline: *Biological Abstracts, Computer Abstracts,* and so on. For some current research, you might consult abstracts of doctoral dissertations in *Dissertation Abstracts International.* Abstracts are increasingly searchable by computer.

Access tools for U.S. government publications

The federal government publishes maps, periodicals, books, pamphlets, manuals, monographs, annual reports, research reports, and other information, often searchable by computer. Examples include *Electromagnetic Fields in Your Environment, Major Oil and Gas Fields of the Free World,* and *Journal of Research of the National Bureau of Standards.* Your best bet for tapping these complex resources is to request assistance from the librarian in charge of government documents. Listed below are the basic access tools for documents issued by or published at government expense, as well as for many privately sponsored documents.

- *The Monthly Catalog of the United States Government:* The major pathway to government publications and reports.
- *Government Reports Announcements & Index:* A listing (with summaries) of more than 1 million federally sponsored research reports published and patents issued since 1964.
- *The Statistical Abstract of the United States:* Updated yearly, it offers an array of statistics on population, health, employment, and the like. It can be accessed via the Web. CD-ROM versions are available beginning with the 1997 edition.

Many unpublished documents are available under the Freedom of Information Act (FOIA). The FOIA grants public access to all federal agency records except for classified documents, trade secrets, certain law enforcement files, records protected by personal privacy law, and the like. Contact the agency that would hold the records you seek: for workplace accident reports, the Department of Labor; for industrial pollution records, the Environmental Protection Agency; and so on. Government information is increasingly posted to the Internet.

Microforms

Microform technology allows vast quantities of printed information to be stored on microfilm or microfiche. This material is read on machines that magnify the reduced image.

Review Checklist

Type of Research	Sources
Primary research	Informative interviews
	Surveys and questionnaires
	Public records and organizational publications
	Personal observation and experiments
Internet research	Usenet news
	Listservs
	Electronic magazines (zines)
	Email inquiries
	The Web
Other electronic research tools	Compact discs
	Online retrieval services
	Card catalog
Hard copy research	Bibliographies
	Encyclopedias
	Dictionaries
	Handbooks
	Almanacs
	Directories
	Guides to literature
	Indexes
	Abstracts
	Access tools for U.S. government publications
	Microforms

Exercises

1. Locate a research article from the past four years (not one that is too recent). Using a citation index, track down the specific publications in which the original material has been cited. If your article is from a scientific or technical field, try the *Science Citation Index*. Based on the number and type of citations to the original article, what is your opinion about the importance of the original article's findings?

2. Locate an expert in the field for your major or an upcoming project (such as a long report). Arrange to interview that person. You may wish to make your initial contact via email. Follow the tips on page 52. In groups of three or four, discuss your interview experience and your findings with your classmates.

3. FOCUS ON ▸WRITING. Identify two major indexes to locate research articles for your field or topic. One source should be a traditional periodical index and the other should be an Internet search engine. Locate a recent article on a specific topic (for example, privacy laws in your state) and write a short summary.

4. FOCUS ON ▸WRITING. Your school is reviewing each department and each student organization on campus, with an eye toward expanding the most promising and popular programs. Therefore, each department and organization is expected to make a case for additional funding and support. Trace the history of your major department (including periods of growth and decline, job prospects for graduates, or other trends), its contribution to your school's overall mission, and its outlook for the coming decade. Or do the equivalent for a student organization to which you belong. Prepare a 2-page memo to the student and faculty senates that argues—on the basis of persuasive evidence—for added funding. NOTE: This assignment will rely largely on primary research.

The Collaboration Window

Divide into small groups and prepare a comparative evaluation of literature search media. Each group member will select one of the resources listed below and create an individual bibliography (listing at least 12 recent and relevant works on a specific topic of interest selected by the group):

- conventional print media
- electronic catalogs
- CD-ROM services
- a commercial database service such as DIALOG
- the Internet and World Wide Web
- an electronic consortium of libraries, if applicable

After carefully recording the findings and keeping track of the time spent in each search, compare the ease of searching and quality of results obtained from each type of search on your group's selected topic. Which medium yielded the most current sources? Which provided abstracts and full texts as well as bibliographic data? Which consumed the most time? Which provided the most dependable sources? The most diverse or varied sources? Which cost the most to use? Finally, which yielded the greatest depth of resources? Prepare a report and present your findings to the class.

The Global Window

Focus on ▥➡Writing. To study the effects of prolonged space travel on humans as well as on plants and animals, American astronauts lived with Russian cosmonauts on the Russian space station Mir from 1995 to 1999. This was the first phase of the $60 billion project to build the International Space Station. This project will require collaboration among experts from 15 countries, as well as support from political, social, and industrial leaders of these countries.

In order for such a complex undertaking to succeed, people from different cultures have to communicate effectively and sensitively, creating goodwill and cooperation. Before your organization, NASA, begins work in earnest with a particular country, your coworkers will need to develop a substantial degree of cultural awareness.

Your assignment is to select a country and to research that culture's behaviors, attitudes, values, and social system in terms of how these variables influence the culture's communication preferences and expectations. What should your NASA colleagues know about this culture in order to communicate effectively and diplomatically about the proposals regarding the space station? For example:

- Some cultures hesitate to debate, criticize, or express disagreement.

- Some cultures observe special formalities in communicating (say, expressions of concern for one's family).

- Some cultures consider the source of a message as important as its content. Establishing trust and building a relationship may be essential preludes to any professional activity.

- Cultures differ in their attitudes toward big business, technology, competition, or women in the workplace. They might value delayed gratification more than immediate reward, stability more than progress, time more than profit, politeness more than candor, age more than youth.

Click on This

Because the Web is full of information from a variety of sources, students and researchers alike may have a difficult time determining how to judge the validity of information on a given Web site. In groups of three or four, make a list of the items you generally take into consideration when evaluating a Web site for research. For example, do you consider good screen design a sign of credibility? What about recent information or links that work? After making your list, do an experiment: Use a search engine and locate Web sites for a particular topic (such as global warming, genetically modified organisms, or computer privacy). Select one site and evaluate its credibility as a source for research information. Add any new criteria to your list.

Many libraries, schools, teachers, and others have created such lists (of how to judge a Web site's credibility). Visit these sites to see how your list compares. A few are located at:

www.webcredibility.org   The Web Credibility Project at the Stanford Persuasive Technology Lab. The goal of this project is to evaluate the factors that users take into account when considering if a Web site is credible.

www2.widener.edu/Wolfgram-Memorial-Library/inform.htm   "How to Recognize an Informational Web Page" from the Wolfgram Memorial Library, Widener University (Chester, PA).

www.library.cornell.edu/okuref/research/webeval.html   "Evaluating Web Sites: Criteria and Tools" from Cornell University.

www.unc.edu/cit/guides/irg–49.html   "Evaluating Web Sites for Educational Uses: Bibliography and Checklist" from the Center for Instructional Technology at the University of North Carolina at Chapel Hill.

CHAPTER **5**

Technical Communication in a Digital World

Communicating in digital space

Technical communication principles still apply

Designing information for the new media

Writing issues

Design issues

Technical issues

Online documentation and interface design

Corresponding over the wires: email conventions

Working on the wires

Presentation software

Review checklist

Exercises

The collaboration window

The global window

Click on this

Communicating in digital space

Computer technologies are often touted as the answer to communication problems. Much of what formerly took hours to accomplish by hand takes only seconds on a computer. For example, creating a pie chart or line graph once required careful hand drawing using pen, ink, and colored overlays. Now, clicking on a button in any spreadsheet program and creating a chart, graph, or other visual display takes only seconds. Similarly, contacting people by telephone used to be difficult at times, but voice mail and pagers simplify this task.

But these technologies by themselves do not guarantee quality communication. In fact, the information overload that often results from the use of so many technologies can make communication more, not less, confusing. However, by understanding the unique features of digital communication, you can use these technologies to create quality technical communication.

Technical communication principles still apply

All the elements discussed in Chapter 1 (accessibility, usability, and relevance) apply to electronic communication as well. In addition, there are many unique aspects of communication in the digital realm.

Designing information for the new media

Web sites, digital television, voice activation systems, and other technologies are often called the "new media" because they blend characteristics of many traditional media types. For example, print newspapers are different from television news. Radio ads are distinct from brochures. Software manuals are distinct from CD-ROMs. Yet the trend today is toward "convergence," meaning that lines between these forms of communication are becoming blurred. Online newspapers blur the distinctions between a newspaper and an interactive Web site, for example. As more and more of this convergence occurs, technical communicators face information design issues unique to the new media.

For instance, if you are a technical writer at a software company, you may be asked not only to research and write the information for a user's guide, but also to design this information for a Web site or an online help screen. In many organizations, these tasks are done by different communication specialists. Companies may have employees who focus solely on writing, while others focus on Web design. Even so, each communication specialist must know the principles involved in designing for new media.

Although new media can refer to many types of information, this section focuses on Web page design. When designing for the Web, you should consider three areas: writing issues, design issues, and technical issues.

Writing issues

Like a paper document, writing for Web sites must conform to the principles of effective technical communication. But writing for Web sites involves additional audience considerations.

Addressing the needs of impatient readers. Web page readers generally have less patience with onscreen material than with hard copy; one Web expert claims that "[w]hen visiting a new site, users often give up on it before its main page has fully downloaded" (Rosenfeld & Morville, 1998, p. 8). Computers are associated with speed, and people are impatient with long blocks of online information. Also, the flickering light patterns of computer screens tire the eyes, causing people to resist reading long passages of text online.

To counter reader impatience, you should write text in a *chunked* format, breaking down the text from one long paragraph into shorter passages that are easy to access and quick to read. Chunking is also used in paper documentation (see Chapter 3) but is especially important for Web documents.

How you chunk the information depends on the way your audience intends to interact with the site. One group of experts recommends that chunking of information "be flexible and consistent with common sense, logical organization, and convenience. Let the nature of the content suggest how it should be subdivided and organized" (Lynch & Horton, 1999, p. 25).

Imagine, for instance, that you are writing text for a Web site intended to explain a medical procedure to patients. After performing an audience analysis, you decide to organize this information by questions users might ask: "How long will the surgery last?" or "When will I be able to return to work?" You would chunk your information into these categories, possibly making each an individual link on the Web site.

Companies increasingly realize that chunked information is cost-effective and strive to create single sources of information that can be reused in various formats. A technical communicator may write the information once, store it in a database, and use the same chunks (or "modules") to create a manual, a brochure, a Web site, and a CD-ROM.

Addressing the needs of nonlinear readers. People do not read Web sites in a linear fashion. A typical Web site displays information in a *hypertext format*, allowing users to jump around, moving from link to link, often out of the originally intended sequence. Each chunk of text on a Web page must make sense, regardless of the order in which it is read.

Addressing the needs of a diverse audience. A Web page's audience may be far broader than originally intended. Unless you are designing an Intranet site (available only within a specific company or organization), your Web site can be accessed by almost anyone worldwide. Some Web sites, such as a site for patients seeking specific medical information, begin with a brief description of the intended audience. To accommodate a global audience, take the following steps:

- Provide links to information that might be highly technical or require further explanation.
- Provide a feedback mechanism (email address, Web form) for people who have questions.
- Write in short sentences, not only for ease of reading, but also for easy translation to Web pages in different languages.

Design issues

Web pages are designed using *hypertext markup language,* or HTML. This HTML code appears as inserted commands, or "tags," which, when read by a Web browser (such as *Netscape*), cause formatting features to appear on the screen. For example, a line of text tagged as follows:

    <c>Privacy and Ecommerce</c>

would appear as bold () and centered (<c>) when viewed through a browser:

Privacy and Ecommerce

The slash tags at the end of the text (and </c>)tell the browser to turn off the boldface and centering features until these codes are used again.

In the Web's early days, HTML coding was done by experts. Today, it is easy to write a Web page with little or no knowledge of HTML. You can use Web editing software or even most word-processing programs to create a Web page, much as you would create a word-processing document. The software then translates your document into HTML code.

Even though Web sites are technically easy to produce, effective design is no simple matter. Many sites created by novices are poorly designed and hard to use. Web design is the subject of many books, classes, and workshops. Listed below are some basic design features to consider.

Organization. One common source of frustration for Web users is the "can't find it" problem (Rosenfeld & Morville, 1998, p. 4). Most of us have visited Web sites that have no links, no overview of the site's content, or no apparent logical organization. A well-designed Web site is clearly organized so that users can follow the information flow (see Figure 5.1). The first page should be organized top-down, providing an overview of the entire site and links to different chunks of information. Links should be structured according to their "hierarchy of importance," moving from the more general information chunks to the more specific (Lynch & Horton, 1999, p. 25).

Figure 5.1 ■ A well-organized Web page. The home page for the Public Broadcasting Service is well organized. Main categories of information are clearly marked along the left-hand side, and special features, highlighted with graphics, are shown along the right. There is also a link at the bottom where users can connect to many other PBS-affiliated sites. Even with all this information, the site is easy to follow.
Source: PBS, www.pbs.org

Typography. Typefaces on a computer screen show up less clearly and with less resolution than on a printed page. Also, predicting exactly how a typeface will appear on a given computer screen is almost impossible, since different browsers present fonts differently. Lynch and Horton (1999) offer these recommendations for type design on the Web:

- Present a clear size contrast between one font and another.
- Give attention to the use of capital and lowercase letters. Don't use all caps, which are hard to read.
- Use a serif font for the body text and sans serif for display text.
- Use typefaces designed specifically for reading on computer screens, such as Georgia or Verdana. (pp. 79–90)

Notice how most of these recommendations are embodied in Figure 5.2.

Line length. Computer screens curve at the edges, making long lines of text hard to follow. Never simply "dump" a page from print text onto a Web site. Instead,

FɪɢᴜʀE 5.2 ■ A typographically effective Web page. This Web site uses typography effectively for the screen. The attention-grabbing style used for the headline is clear and readable. For the body type, sans serif (block) typefaces are easy to read on the screen.
Source: www.technicalinspirations.com

chunk the text (described earlier) into smaller units. Make lines of text at least ¹/₂ inch narrower than the full screen width (Figure 5.3a and b). Also, text set ragged right is easier to read than text set justified right.

Appearance. A well-designed, visually attractive Web site combines the information power of both printed text and visual information (such as a photograph, illustration, or live video). Color can increase visual appeal, but don't mix too many colors. Select a few colors that complement each other well, and keep your choices simple. Use colors that are appropriate for your audience: The school colors, for example, might make an attractive background for a university Web site.

Technical issues

Web sites must be *accessible*. For example, a Web site that contains photographs and illustrations may be visually attractive but hard to access for users whose computers do not have sufficient memory or processing power. In order to

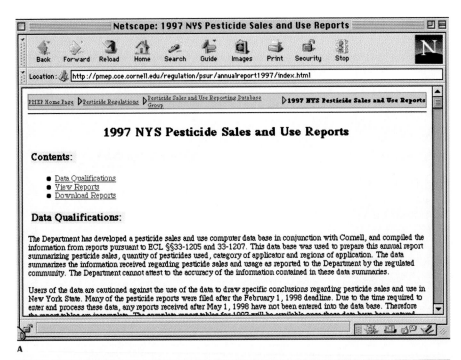

A

b

Figure 5.3a and b ■ Figure 5.3a illustrates lines that are too long for a computer screen. Although the entire text is visible, lines that take up the whole screen are hard to read. Figure 5.3b uses a shorter line length, which is easier on the eye.

Sources: Figure 5.3a, Pesticide Management Education Program at Cornell University, Ithaca, NY. Figure 5.3b, B. L. Travel, Poultney, Vermont.

make your site as accessible as possible, learn all you can about the computer use of the majority of your audience. Work with your company's technical support staff to use *standard file formats* and design a site that is suited for the browser and computers used by the audience. Web sites for a global audience should be compatible with the current industry standard browsers (*Netscape* or *Microsoft Explorer,* for example). Make sure all links work correctly, and avoid the "gratuitous use of bells and whistles" (Rosenfeld & Morville, 1998, p. 5). Finally, test the site to be sure it is *functional* and *usable*. Most organizations have staff who work specifically on the technical aspects of Web sites and can assist you with making sure the technical design of your site is appropriate for your audience.

ONLINE dOCUMENTATION ANd INTERFACE dESIGN

Online documentation is documentation that is either delivered to the user via a CD-ROM or built into the software itself. If you have ever pressed the "help" key in your word-processing software and read the help information on your screen, you have accessed online documentation.

Online documentation

Online documentation (Figure 5.4) is very popular in the software industry, because it saves time and money in printing paper documentation and manuals. Also, some online documentation is *context-sensitive*: If a spreadsheet user makes a mistake, for example, the software will recognize the mistake and direct the user to the appropriate help screen. This process is much easier for the user than sifting through pages and pages of a paper manual.

Like Web pages, online information should be written in well-organized chunks and should never be paper documentation dumped into an electronic format. To create effective online documentation, one expert suggests listing all the tasks people perform with the paper document, then deciding whether each task is possible in the online version. If people rely on an index to look up information in a specific manual or report, an index might also be needed in the equivalent online document (Horton, 1990).

You can use special software, such as *RoboHelp* or *Doc-to-Help*, to convert textual material into online help files.

Interface design

The interface is the part of a software application you see on the screen. When you run a word-processing program, for example, you see a menu bar, buttons to click on, and a background screen. While the heart of the program, computer code, runs silently in the background, the interface, the clickable icons and menus, lets users interact with the machine.

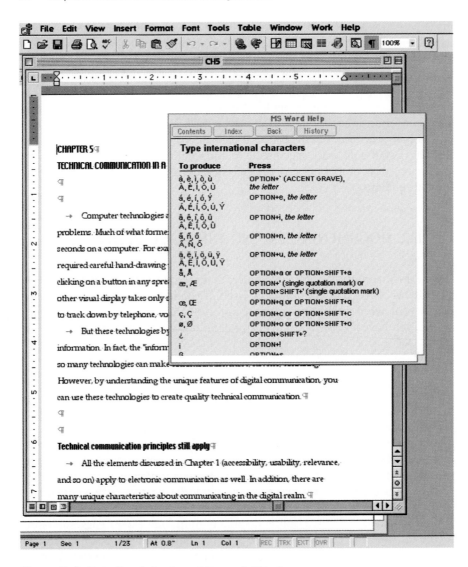

Figure 5.4 ■ Online help from *Microsoft Word*.

Source: Reprinted by permission from Microsoft Corporation.

Software interfaces must be well designed: Menu commands must be consistent, visual images must appear in logical places, and features must perform in ways that make sense to the user. Items must be spelled correctly, and screens must appear in the proper order.

Interface design is such a complex subject that many computer science departments offer degrees in this area. Technical communicators often work on the layout, functionality, and wording of software interfaces, and some go on to become interface designers.

Corresponding over the wires: email conventions

Features that make email attractive for workplace communication include:

- **Asynchronicity.** Asynchronous communication does not take place in real time. Unlike a face-to-face conversation, in which two or more people must be physically present, email allows you to communicate with someone any time, day or night. You can send a message at 2:00 a.m., if that suits your schedule, and the recipient can read it when she arrives at the office at 9:00 a.m.
- **Electronic "paper trail."** Unlike phone calls or face-to-face conversations, email messages can be stored and saved for future reference. It is also possible to cut and paste material from an email message into another document.
- **Easy forwarding.** Email messages can be forwarded to others in a single keystroke, thus simplifying the distribution of a message to multiple recipients.
- **Attachments.** Most email programs enable you to send attachments (Figure 5.5) of entire documents with their original formatting. Word-processing files—of your resume, a formatted report, or any other document—as well as spreadsheet, sound, or graphics files can easily be attached to most email messages and read by the recipient. Most popular email programs make attachments between PCs and Macintosh computers interchangeable.

Figure 5.5 ■ **An email message with attachment.** This email message was sent in *Microsoft Outlook for Windows.* The paper clip in the right blue area indicates an attachment.

Source: Reprinted by permission from Microsoft Corporation.

Technical communicators commonly share drafts of material via email or co-author entire reports or other documents by collaborating through email.

The heavy use of email has resulted in new conventions for spelling, phrasing, and expectations of how quickly people will respond. Although many people have tried to outline "netiquette" rules, these continue to evolve. When communicating via email, consider the following issues:

- **Oral or written.** At first glance, email resembles a written document. The standard email format resembles that of a paper memo, with "To," "From," "Date," and "Re:" fields. Yet email tends to be informal and conversational. Even writers who are extremely careful with traditional written correspondence sometimes ignore spelling or grammar, instead writing an email message much as they might speak it. Spell check your email messages whenever possible, and avoid using all capital letters (ALL CAPS in email is considered shouting).

- **Speed and reach.** As with Web sites, people have a short attention span when it comes to reading email. Email is considered a *speedy* medium, because people read messages quickly. Don't send extremely long messages or huge attachments unless you have warned the receiver in advance. Also, remember that accuracy is essential in technical communication. Email seems to inspire writers to send information quickly, without proofreading for content. Be sure to check the content of your message carefully.

  Reach refers to the idea that email messages can be seen by vast audiences across great distances. You never know who will read your message. A message intended only for your manager may accidentally get sent to the entire department. Or your manager may forward a message, intact, to others in the organization—and that message may be the only impression others have of you. In most organizations, email messages belong to the company, not you. Never post an email message that would be a problem for you if it got posted to people other than your immediate intended audience.

- **Flaming.** Some people use email to express anger and to personally attack others. Researchers speculate that this behavior, called *flaming,* has various causes, including the speed at which people post messages and the fact that email writers can hide behind the screen, thus avoiding the repercussions of face-to-face rudeness. If your email exchange becomes difficult, confusing, or angry, consider calling a meeting or making some phone calls to resolve the situation.

- **Hierarchy.** Email often lets you bypass "pecking orders" and send messages directly to someone you might never reach by phone. Because email tends to inspire informality, writers often forget they are not writing to an old college roommate but to a manager, a respected scientist, or an important author. Even with email, politeness is always important; for example, if you don't know the recipient of your message, begin with a salutation—"Dear Dr. Jones"—not with an informal "Hi!" or with no salutation at all.

- **Gender in cyberspace.** Research suggests that females may participate less or be overwhelmed by male opinions on the Internet. While there is no conclusive evidence that this is always the case, researchers know that young girls often get less time at the computer, at home and in school, and that in workplace situations, men often dominate the conversations during meetings. Make sure everyone involved in an electronic discussion (a series of email messages about a project, for example) has an equal say. If you find that someone (of either gender) is not speaking up, you may wish to encourage that person to contribute. Or, if someone on the list is dominating the discussion, privately email that person and ask him or her to allow for other voices.

- **Spelling.** Writers generally proofread paper memos, but because email is often written and sent rapidly, less attention is paid to spelling, even though many email programs contain spell checkers. Take a few minutes to spell check email messages, especially those written for business.

WORKING ON THE WIRES

Increasing numbers of employees work from home, via a computer and modem and with the help of voice mail and a fax machine. This *telecommuting* requires great attention to detail. Information can be misconstrued, lost, impossible to download, or improperly posted when sent via computer. Faxes do not always arrive. Voice mail messages can be accidentally deleted. Any projects conducted via telecommuting technologies must be checked carefully.

Some organizations offer ways for employees to "meet" across vast distances in real-time settings. Software called *computer-supported cooperative work systems,* for example, allows people at workstations across the world to connect, send messages, and participate in live electronic conversation.

Employees often take *distance education* courses via the Web or interactive television. Technical professionals often take such courses to enhance their skills or pursue advanced degrees. Also, technical communicators often help design such courses.

PRESENTATION SOFTWARE

Presentation software (such as *PowerPoint,* discussed in Chapter 10) allows you to enter text, images, and animation and turn this information into presentation slides, which can be displayed via the computer or printed out as overhead transparencies and handouts. Although this software can spice up an oral presentation, offering backgrounds, color, and transitions, it is no substitute for a well-organized and well-prepared presentation. Again, the lure of the technology should never eclipse the principles of effective technical communication.

Review Checklist

Issues of Web Design

Issue	Characteristics	Guidelines
Writing	Text	• Chunk information for easy access and quick reading. • Consider audience needs. • Create chunks (modules) for multiple purposes.
	Nonlinear readers (hypertext)	• Write each chunk to be understandable regardless of order or context.
	Broad audience	• Define all terms and provide necessary links. • Provide a feedback and question mechanism. • Write short sentences to facilitate reading and translation.
Design	Organization	• Use top-down organization • Structure links by hierarchy of importance
	Typography	• Follow principles of good type design: Use clear contrast between fonts; avoid using all capital letters; use a serif font for body text and sans serif for display text. • Use fonts designed for computer screen reading (Georgia, Verdana).
	Line length	• Chunk information into small units. • Use short text lines. • Avoid right-justified text.
	Appearance	• Focus on cultural and organizational considerations • Avoid mixing too many colors, and use colors appropriate to the audience
Technical features	Accessibility	• Consider the technical capabilities of audience's computers. • Use standard file formats. • Design for industry standard browsers. • Ensure that all links are correct. • Avoid gratuitous use of bells and whistles.
	Usability	• Test site to ensure it is functional and usable.

Characteristics of Using Email

Characteristic	Comments
Asynchronicity	• Communicate any time, day or night.
Electronic paper trail	• Store messages for future reference. • Cut and paste contents into other documents.
Forwarding	• Distribute messages easily to others.
Attachments	• Transmit word-processing, sound, graphics, and spreadsheet files as attachments.
Tendency to resemble spoken communication	• Avoid the temptation to ignore spelling and grammar conventions.
Speed and reach	• Avoid sending extremely long messages. • Check for accuracy. • Many people besides the intended recipients may read messages.
Flaming	• Be alert to difficult, confusing, or angry exchanges and intervene.
Hierarchy	• With access to those traditionally unavailable, remember to be polite.
Gender	• Ensure that everyone in email exchanges participates equally in the discussion.
Spelling	• Messages with correct spelling have more credibility than those that are sloppy.

 EXERCISES

1. FOCUS ON ⟶ WRITING. With a classmate, locate one or two Web sites you might use as research sources for a project. Using the guidelines in this chapter, assess these Web sites for the quality of their writing, design, and technical features. How much does the visual attractiveness or choice of fonts affect your initial view of a Web site? Write up your findings in an email message to your instructor. Include the Web addresses for each of the sites you used.

2. FOCUS ON ⟶ WRITING. Interview a professional in your field, and ask how this person's job is affected by new technologies. For example, you might ask how much of his or her work is conducted via email. Write a memo to your instructor and fellow students explaining the amount of time a professional might anticipate spending on a daily basis using communication technology.

3. FOCUS ON ▶WRITING. Using word-processing software, compare online help information to the information in the paper manual. Notice in particular the use of chunking. Is the online information organized differently from the paper? Which is easier to use, and in which case can you find the topic you are searching for more quickly and more accurately? Write a short report comparing the two types of information. Include both online and hard copy examples.

The Collaboration Window

Technical professionals often find they must collaborate on projects, and many of these professionals use technology to enhance their collaborations. In class, form teams of three or four people. Assume that you are all located in different parts of the world and must collaborate on a report on a topic that you choose. After determining an audience, purpose, and scope for your report, create a first draft using the computer to share information. You may wish to set up a Web site where group members can post their sections of the paper or use email attachments to pass the report back and forth.

As you work on this project, ask each team member to keep a log of any technical issues that arise. For example, if you use the Web to post information, can all team members access the Web site? If you use email attachments, are all members able to open the same file types? Combine your log into one list, and based on this list, draft a set of guidelines for collaborating via the Internet. Share these in a brief oral presentation to class.

The Global Window

Find Web sites that are in different languages or based in different countries. Besides differences in language, note any other differences between non-U.S. Web sites and those that originate in the United States. Some features you might look for include:

- **Use of color.** Different cultures often associate unique meanings with certain colors. For example, the color red may be used in India to mean procreation or life, while red in the United States is often associated with danger or warnings (Hoft, 1995, p. 267).

- **Issues of privacy.** U.S. sites often give out *cookies* (files sent to your computer that give Web site providers information about you), but other countries, such as Germany, take a different approach, allowing no personal information to be used without an individual's permission. Note if any of the sites mention their privacy policy in this regard.

- **Date formats.** The typical U.S. format for dates is month, date, year (MDY), as in December 6, 2000 or 12/6/00, while the French or German format is date, month, year (DMY), as in 9 Dezember 2000 or 9.12.00 (Hoft, 1995, p. 232).

Keep track of what you find, and present your findings in class. If you can, create a class Web site that describes your findings and links to interesting sites on intercultural communication.

Click on This

- The *Internet Scout Report,* published by the University of Wisconsin-Madison's Department of Computer Sciences, seeks out high-quality sites on the Web. Check out the *Scout Report* at scout.cs.wisc.edu/scout. Also, notice that the *Scout Report* publishes information on how items are selected. You can read about this at scout.cs.wisc.edu/scout/report/criteria.html. In class, discuss criteria for Web site credibility.

- Effective technical communication should be accessible, but the Web may not always be accessible to people with disabilities. A set of guidelines, released by the World Wide Web Consortium (W3C), explains how to make Web content accessible to people with disabilities and to other users as well. The document offers 14 Web Content Accessibility Guidelines. Check them out at www.w3.org/TR/1999/WAI-WEBCONTENT–19990505.

- The Yale CAIM Style Guide (also available in print; Lynch & Horton, 1999) is an excellent source of information about Web design. It is available at info.med.yale.edu/caim/manual.

CHAPTER **6**

Ethical Issues in Technical Communication

Ethics, technology, and communication

Case examples of ethical issues in technical communication

Types of ethical choices

What is legal is not always ethical

Additional ways in which actions can be unethical

Types of technical communication affected by ethical issues

Responding to ethical situations

Review checklist

Exercises

The collaboration window

The global window

Click on this

ETHICS, TECHNOLOGY, AND COMMUNICATION

Technical communication does not occur in a void. It happens in the world of human beings, politics, and social conditions; a world in which we regularly face ethical dilemmas that pit our sense of what is right against a decision that may be more efficient, profitable, or better for the company.

Ethical questions often revolve around topics related to technology. For example, a new computer chip that secretly collects personal information about a person's Web surfing habits presents a privacy dilemma. Should users be allowed to choose whether to give out this information? Some would say yes, but in the United States, few laws address personal privacy at this level. So, the decision becomes an ethical one. The communication about this product (a press release announcing it or a user's manual that accompanies the computer) plays a central role in this ethical dilemma. Should the technical writer include this information, exclude it altogether, or de-emphasize it by using a small font?

These are not simple questions. Taking an ethical stance requires a personal decision on your part as to how to balance your ethical and moral beliefs with the realities of the job. It requires you to consider the effects of your decisions on the users of your product, on your company, on society at large, and on your job. Sometimes, standing your ground on an ethical issue may mean losing your job or suffering retaliation from coworkers.

CASE EXAMPLES OF ETHICAL ISSUES IN TECHNICAL COMMUNICATION

The disaster at the Three Mile Island nuclear power plant, the explosion of the Space Shuttle *Challenger,* and the decision to market the Pentium III chip help illustrate the relationship between ethics and technical communication.

Three Mile Island[1]

The case. On March 28, 1979, the Three Mile Island Nuclear Power Plant near Harrisburg, Pennsylvania accidentally leaked radioactive gases through the plant's venting system. Many feared this leak might lead to a runaway fission reaction and total meltdown, which would have been an environmental and social disaster, not only for the immediate area, but also for the entire region and possibly the nation (Miles, 1989, p. 44). At the time, U.S. citizens were questioning the safety of nuclear power, and the Three Mile Island disaster only reinforced these concerns.

[1]This section is adapted from Miles (1989).

The communication situation. In the months that followed, a Presidential Commission's study of the accident revealed a breakdown in communication among engineers at Babcock and Wilcox, the private company that ran the Three Mile Island plant. One particular memo was directly implicated in the accident.

- **The memo.** Roughly 18 months before the Three Mile Island accident, a similar problem had occurred at another nuclear power plant, also operated by Babcock and Wilcox. As a result, a managing engineer wrote to several other managers, addressing serious concerns within the system. Specifically, the memo suggested that "core uncovery" (meltdown) might occur if this problem was not corrected. Yet the writer placed this vital information in the middle of the memo, a place readers often skip over. In part because this memo was not emphatic enough, the problem was overlooked until the Three Mile Island plant's near miss with nuclear disaster.

The space shuttle *Challenger*[2]

The case. On January 28, 1986, the space shuttle *Challenger* exploded 43 seconds after launch, killing all seven crew members. The immediate cause was that two rubber O-ring seals in a booster rocket permitted hot exhaust gases to escape, igniting the adjacent fuel tank. However, the O-ring hazard had been recognized since 1977 and documented by engineers but largely ignored by management. (Managers had claimed that the O-ring system was safe because it was "redundant": each primary O-ring was backed up by a secondary O-ring.)

Moreover, in the final hours, engineers argued against launching because that day's low temperature would drastically increase the danger of both primary and secondary O-rings failing. But, under pressure of deadlines, managers chose to relay only a downplayed version of these warnings to the NASA decision makers who ultimately were to decide on the *Challenger*'s fatal launch.

The communication situation. Various aspects of technical communication were involved in this situation. These included:

- **Organizational role.** Engineers were concerned with safety features, while managers were concerned with making the launch on the date and time it was planned. During meetings, these different points of view often clashed. One engineer was even told to "take off your engineering hat and put on your management hat" (Presidential Commission, 1986, p. 93).

---

[2]This section is adapted from Winsor (1988), Pace (1988), Gouran, Hirokawa & Martz (1986) and Gross & Walzer (1994).

- **Written communication.** Many memos and technical reports circulated during the discussion of whether or not to launch. Writers of these memos had to make choices: Should they emphasize the danger of the situation at the possible expense of losing face with their managers? Should the wording be strong or cautious? One engineer decided to word his memo in the strongest possible terms, but the memo never reached the top-level decision makers at NASA.

The Pentium III Chip[3]

The case. In early 1999, the makers of the personal computer chip called the Pentium announced a new chip, the Pentium III. While the manufacturer and computer makers heralded the chip, privacy advocates were concerned. The chip contained a unique serial number that would automatically activate when the computer was in use. Intel, the chip's manufacturer, argued that this serial number would be useful for ecommerce, because it would give Web sites a unique method to identify a user each time that person logged on. But privacy groups were concerned that the chip would invade personal privacy—users would be giving up information about their Web activity without consent, and it might be possible for an unauthorized user to access the Pentium serial number and make credit card purchases on the real user's account.

The communication situation. As users became increasingly concerned about the privacy implications of the Pentium III chip, a communication situation developed around the following issues:

- **Email and Internet communication.** The Internet became the main discussion forum for the Pentium III debate. On the one hand, corporate communication from Intel and other computer companies claimed that the chip would pose no privacy problems and that, in fact, similar identifying information was already being collected when users logged on to Web sites. Advocates also suggested that the chip, with its unique ID number, could help track down a stolen computer. On the other side, privacy and civil liberties organizations were alarmed that users had no way of turning off the chip and felt that the chip was part of a broader problem of technologies that collect personal information. These discussions took place on Web sites, via discussion lists, and through email that was passed back and forth across the Internet. Some of the information was true; some was inaccurate. Anyone interested in the debate had to sort through this vast amount of information from different perspectives: Intel's information, while technically accurate, was biased toward their point of view; privacy advocates were writing from their perspective, and individuals were often mixing factual information with exaggeration.

[3]This section is based on information from EPIC (2000) and ZDNET (2000).

- **Media reports.** Press releases, newspaper articles, and television and radio reports were equally confusing. Some indicated that the Pentium III chip was a privacy invasion; other sources were more supportive. This information often made its way back to the Internet discussions, causing even more confusion. In the end, the uproar over the chip resulted in Intel's creation of a mechanism that allows consumers to turn the identifying feature on or off. (But privacy advocates were still not convinced that this was a good solution.)

The Pentium III case may not be as clear-cut as the Three Mile Island and Space Shuttle *Challenger* ones in that there is no single "smoking gun" (such as the memos in those cases). But what this case illustrates is that controversial technologies raise ethical issues that are not always easy to sort out. A communicator employed at Intel who disagreed with the Pentium III's approach to collecting private information would have to weigh his or her ethical stance against the interests of the company. And citizens who wanted to learn more about the chip found mixed information at best.

Virtually all areas of science and technology are involved in issues of communication and ethics:

- *Medical technologies,* such as genetic testing, raise questions about personal privacy and medical insurance.
- *Banking and retail operations,* which increasingly collect personal information on consumers, raise concerns about how this information is used and who has access to it.
- *Environmental pollutants,* such as pesticides or smoke stack output, raise serious questions about the long-term health of the planet.

In your own communication, you will often face ethical decisions—about how much information to leave in, how much to leave out, how to word an issue, or how to shape the information for users and consumers. In the end, communication is never neutral. It always comes with some type of consequence.

Types of Ethical Choices

Throughout time people have tried to define universal principles of ethics to provide a basis for ethical decision making. Here are some of these theories.

Kant's categorical imperative

Immanuel Kant (1724–1804) argued that certain ethical situations dictate certain actions. Kant suggested that "codes of conduct and morality must be arrived at through reason and be universally applicable to all societal environments at all times." Kant emphasized the individual's responsibility and the intention of the act, not its consequence (Fink, 1988, p. 7).

Utilitarianism

Associated with John Stuart Mill (1806–1873), this ethical principle asserts that "ethical conduct should aim at general well-being, creating the greatest happiness for the greatest number of people." Unlike Kant, Mill argued that the outcome, not just the intention of an act, determined how people judge the ethics of that behavior (Fink, 1988, p. 7).

Ethical relativism

Taken to its extreme, ethical relativism suggests that any act may be ethical, depending on the particular ethical, religious, and cultural stance of the individual or group. A more moderate approach to relativism would suggest that acts need to be considered not against some fixed set of standards, such as Kant's notion, but in the context of the culture and individual circumstance. More recently, some would argue, U.S. culture in particular has gravitated toward a relativistic position regarding ethical behavior.

Such broad ethical principles, however, rarely provide sufficient guidance for the countless ethical decisions technical communicators face today. Kant's codes of ethics or Mills's utilitarianism might have been useful in their given periods of history, but today, most philosophers and ethicists agree that in a complex world, it is more effective to consider the particular situation and to develop standards appropriate for that situation.

For technical communication, you might consider an approach based on *reasonable criteria* (standards that most people consider acceptable), which take the form of obligations, ideals, and consequences (Ruggiero, 1998, pp. 33–34; Christians Tackler, Rotzoll, & McKee, 1978, pp. 17–18). *Obligations* are the responsibilities you have to everyone involved:

- Obligation to yourself to act in your own best interest and according to good conscience
- Obligation to stand by the clients and customers to whom you are bound by contract—and who pay the bills
- Obligation to your company to advance its goals, respect its policies, protect confidential information, and expose misconduct that would harm the organization
- Obligation to coworkers to promote their safety and well-being
- Obligation to the community to preserve the local economy, welfare, and quality of life
- Obligation to society to consider the national and global impact of your actions

When the interests of these parties conflict—as they often do—you have to decide where your primary obligations lie.

Ideals are the values that you believe in or stand for: loyalty, friendship, compassion, dignity, fairness, and whatever qualities make you who you are.

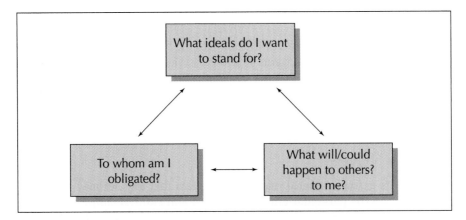

Fiqure 6.1 ■ Reasonable criteria for ethical judgment.

Consequences are the beneficial, or harmful, results of actions. Consequences may be immediate or delayed, intentional or unintentional, obvious or subtle. Some consequences are easy to predict, some are difficult, some are impossible. Figure 6.1 depicts the relations among these three criteria.

Whɑt is leqɑl is not ɑlwɑys Ethicɑl

Based on the above criteria, you can see that just because certain actions are legal does not necessarily mean that they are ethical. Copyright, plagiarism, and privacy examples illustrate this point.

Copyright

Chapter 7 describes certain circumstances under which you may use copyrighted material without the copyright holder's permission. Under the fair use doctrine, for example, you are allowed to use certain materials for educational purposes. But even though your use of the material may be legal, in some circumstances you should question the ethical implications of that use. Consider the following scenario:

> *For a class called The Language of Cyberspace, you are assigned to do a presentation on the ways in which people form communities on the Web. You locate a Web site for people who are struggling with a certain medical condition, and you notice how the users of this site seem to have formed a community among themselves, discussing specific personal aspects of their condition and seeking advice from each other. You make overhead transparencies of some of these Web discussions, telling yourself that you don't need permission because you*

are using only a portion of the material and you are using it for educational purposes. You use these transparencies in a class presentation.

Even though this use is legal, is it ethical? Your obligation to your coworkers (fellow students) and to the community should remind you to consider that the people who logged onto this Web site never expected their names and personal information to be put up on an overhead in front of a class. What if one of the users turns out to be the friend of a classmate, and that classmate never knew about the friend's medical condition?

In this case, you should have considered not only the legal aspects of your decision but also the ethical ones. Your obligation to yourself to act in good conscience might lead you to ask how to avoid causing possible embarrassment to the users of this site, either by selecting a different site for your presentation (one that does not use names) or by changing the names of the users before placing your material on a transparency.

Plagiarism

Plagiarism is using someone else's words and ideas without giving that person proper credit. Even when your use of a source may be perfectly legal, you may still be violating ethical standards if you do not cite the information source.

Assume, for example, that you are writing a class report on genetically modified foods. In your research, you discover a very good paper on the Web. You decide that parts of this paper would complement your report quite nicely. Under copyright and fair use guidelines, you can reproduce portions of this paper without permission. But does this legal standard mean that you can use someone else's material freely, without giving that person credit? Even though it might be legal under fair use guidelines to reprint the material without notifying the copyright holder, using someone else's material or ideas without giving them credit is plagiarism.

Plagiarism is a serious infraction in most settings. Students can be suspended or expelled from school. Researchers can lose their jobs and their standing in the academic community. Most importantly, plagiarism is serious because it violates several of the reasonable criteria for ethical decision making discussed earlier in this chapter. Plagiarism violates your obligation to yourself to be truthful, and it violates your obligation to society to produce fair and accurate information. It also violates your obligation to coworkers, in this case, other students and researchers.

Privacy

Chapter 7 describes privacy issues in technical communication, especially in light of the ability for personal information to be collected over the Internet and the conflicts between U.S. and European approaches to personal privacy.

Since one of your ethical obligations should be to society (to consider the national and global impact of your actions), privacy is high on the list of ethical issues to consider.

If you are designing a Web site, for example, should you create a page that asks users for name, address, and other personal information? If this were a business question, you might automatically say "yes." But since it is an ethical question, you would need to ask about your obligations to society. A privacy statement such as the one in Figure 7.3 might be one way to address both the business and social sides of the question. Another solution is to make sure users have a way to remove their names from your database at any time.

Additional ways in which actions can be unethical

Besides plagiarism or violations of copyright and privacy, your actions can be unethical in other ways. As you read this section of the chapter, consider these situations in relation to the reasonable criteria listed in Figure 6.1. Note that most of these situations relate to workplace issues; Figure 6.2 illustrates how workplace pressures can influence ethical values.

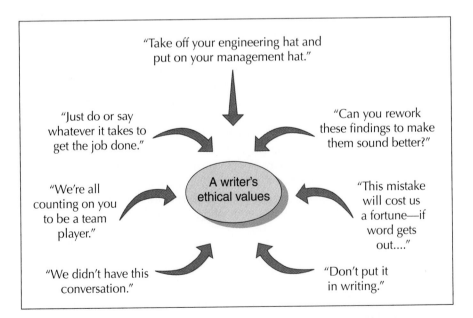

FIGURE **6.2** ■ **How workplace pressures can influence ethical values.**

Yielding to social pressure

Sometimes, you may have to choose between doing what you know is right and doing what your employer or organization expects. Suppose, for example, that just as your company is about to unveil its new pickup truck, your safety engineering team discovers that the reserve gas tanks (installed beneath the truck but outside the frame) may, in rare circumstances, explode on impact from a side collision. You know that this information should be included in the owner's manual or, at a minimum, in a letter to the car dealers, but the company has spent a fortune building this truck and does not want to hear about this problem.

Companies often face contradictory goals of production (producing a product and making money on the profit) and safety (producing a product but spending money to avoid accidents). When production receives first priority, safety concerns may suffer (Wickens, 1992, pp. 434–436). In these circumstances you need to rely on your own ethical standards. In the case of the reserve gas tanks, you may determine that your obligation to society overrides your obligation to yourself or your company. If you make this choice, be prepared to be fired for taking on the company.

Mistaking groupthink for teamwork

Some organizations rely on teamwork and collaboration to get a job done; technical communicators frequently operate as part of a larger team of other writers, editors, designers, engineers, and production specialists. Teamwork is important in these situations, but teamwork should not be confused with *groupthink,* which occurs when group pressure prevents individuals from questioning, criticizing, or "making waves" (Janis, 1972, p. 9). Group members may feel a need to be accepted by the team, often at the expense of making the right decision. To some extent, the *Challenger* case, discussed earlier in this chapter, illustrates groupthink in action. Although several individual engineers had serious concerns about the O-rings in cold temperatures, their concerns were overridden by the sentiments of the group.

Suppressing knowledge the public needs

Pressures to downplay the dangers of technology can result in censorship of important information. For example, high-level employees at the major tobacco companies apparently knew for years about the harmful effects of cigarettes and other nicotine-related products (chewing tobacco, cigars, and so on). Yet lawsuits in the late 1990s proved that many managers and other company decision makers went to great lengths to suppress this information. Should these employees have come forward and admitted what they already knew—that cigarettes cause cancer and other diseases, and that nicotine is very addictive? The answer is "yes" so long as they were prepared to suffer the consequences.

You will need to ask if your obligation to the company takes priority to your greater obligation to your fellow citizens. What about your obligation to yourself to be truthful and to act in good conscience? Again, being aware of your own ethical stance is critical in these situations.

Exaggerating claims about technology

Organizations that have a stake in a particular technology may be especially tempted to exaggerate its benefits, potential, or safety. Assume that you are a technical writer working on the manual and brochure for an uninterrupted power supply, a device that allows computers and other electronic devices to have power even if the main power goes out. Your company manufactures several models of these power supplies. The lowest end model will maintain power for 5 minutes after the main power goes out; the highest end model, for 40 minutes. To emphasize the potential of the product, your manager asks you to only use the 40 minute figure, thereby exaggerating what the other models can do. How would you approach this ethical dilemma? Would you simply do what you were told, or would you find a way to raise the issue with your team and your manager? Your choice will most certainly be affected by your ethical values.

Exploiting cultural differences

Cross-cultural communication carries potential for ethical abuses. Based on their level of business experience or their particular social values, a given culture might be especially vulnerable to manipulation or deception. Some countries, for example, place greater reliance on interpersonal trust than on lawyers or legal wording, and a handshake can be worth more than the fine print of a legal contract. If you know something about a culture's habits or business practices and you use this information at their expense to get a sale or make a profit, you are ignoring your obligations to yourself and your community.

Types of technical communication affected by ethical issues

Certain forms of communication have specific features worth considering in an ethical context.

Graphics

As noted in Chapter 8, graphics are powerful tools for technical communication, and this power can be used in many ways. Graphs, charts, icons, and other images can provide quick, efficient displays of complex information, but they can also be manipulated to distort information. A line graph that does not have labeled axes, for example, might make a financial trend look better than it really is. Other elements of document design, including typography and page layout, can also be used unethically. One study found that technical communicators most frequently based their decisions on a goal-based philosophy; in other words, "the greater the likelihood of deception and the greater the injury to the reader as a consequence of that deception, the more unethical is the design of the document" (Dragga, 1996, pp. 262–263).

Web pages and the internet

The power of a Web page to convey information is obvious, and this topic has been discussed elsewhere in this book. Ethically, you need to consider the speed of the Internet, its global reach, and a Web page's ability to combine sound, color, images, text, and interactivity. These features create the potential for manipulation and distortion. Imagine a Web site for an herbal remedy that some people feel is helpful for anxiety. This herbal remedy may not have FDA approval, and it may have harmful side effects. But a Web site promoting this product could easily, and with very little cost, be set up to look extremely scientific and factual. A fancy logo from quasi-scientific organizations might give the page a sense of professional credibility. Statistics, charts, and links to other Web sites might all give this the appearance of a valid medical site. Yet as a communicator, you need to question the possible outcome for users and the overall risks to society.

Memos

As noted earlier, both the Three Mile Island and Space Shuttle *Challenger* cases involved information conveyed in the form of memos. Memos may seem innocuous enough, but the messages they convey can present serious ethical choices. While the problems at Three Mile Island cannot be placed solely on the shoulders of the memo writer, communicators should be aware that the information in a memo often carries serious ethical implications.

Instructions

Instructions entail a variety of ethical considerations. For example, many instructions contain safety information. Should this information be placed on the first page, or will this deter some consumers from using the product? Again, the answer to this question rests on an ethical decision balancing the safety needs of society against the company's need for profit. Also, if a set of instructions is not tested for usability, technical communicators have no way of knowing if their material is helpful or not. Is it ethical to send out complicated instructions that have never been tested? Why make a user struggle when a test might have revealed errors in the instructions?

Instructions can also easily mislead users in terms of the time required to complete a task. If the average person truly needs two hours to assemble a product, but the instructions say 30 minutes, this information would be unethical.

Reports

Reports must often be kept short in order to fit a certain page format and to be efficient for readers. For conciseness and focus, reports often leave out unimportant information. But who decides what is unimportant, and on what basis? In preparing a report about the environmental impact of your company's pulp and paper manufacturing process, for example, you might be tempted to leave out damaging information, such as the effect of this process on nearby rivers and streams. Or even if you want to include this information, your team or your manager might disagree. So, you face an ethical decision: Should you include everything? What will happen if you do or don't? On what basis should information be left out or included? If you do decide to defy the company, be prepared for anything from a reprimand to the loss of your job.

Proposals

Proposals present specific plans to get something done. A sales proposal, for example, might present your company's plan to design a Web site for a local community organization. Since you are probably bidding against other companies, you want to project the best image and offer the best service for the best price. So you may be tempted to stretch the truth about the time required to complete the job. Is this ethical? On the one hand, such projects almost always exceed the deadline—and if you tell the truth while your competitor lies, you might lose the job. On the other hand, dishonesty is likely to damage your company's (and your) reputation. Other ethical questions raised by proposals include costs and materials to be used.

Oral presentations

In giving an oral presentation, you have the complete attention of your audience. People are listening to you as an expert on the subject, and this face-to-face situation elicits trust between presenter and audience members. Whereas a printed document leaves the writer "invisible," oral presentations create an intimate atmosphere between speaker and listener. In a relative position of power, the speaker can convey accurate information in a fair and balanced manner or can use this trust to manipulate the audience by stirring up emotions without warrant. If your purpose, for example, is to present technical background on a proposed waste incinerator in your community, you should offer the facts as best you can rather than play to the audience's confusion or naiveté.

Responding to ethical situations

To ensure that your communication is ethical, consider the reasonable criteria discussed earlier and shown in Figure 6.1. In addition, many professional organizations have created their own codes of ethics for the particular situations that people in those professions face. For example, electrical engineers can follow the Institute of Electrical and Electronics Engineers (IEEE) code, while nurses can follow the American Nursing Association (ANA) code. Figures 6.3

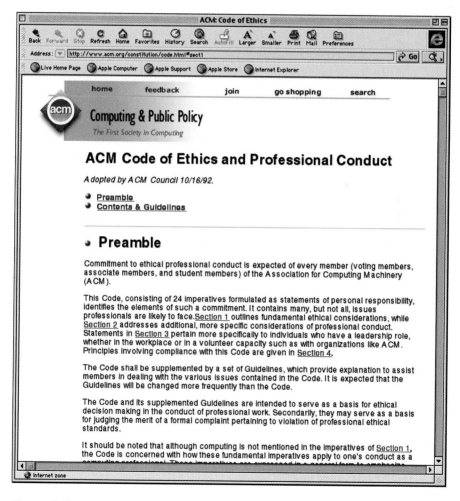

Figure 6.3 ■ **Page 1 of the ACM Code of Ethics and Professional Conduct.**
Source: © 1993, 2000 ACM. Used by permission.

STC Ethical Guidelines for Technical Communicators

AS TECHNICAL COMMUNICATORS, WE OBSERVE the following ethical guidelines in our professional activities. Their purpose is to help us maintain ethical practices.

Legality
We observe the laws and regulations governing our professional activities in the workplace. We meet the terms and obligations of contracts that we undertake. We ensure that all terms of our contractual agreements are consistent with the STC Ethical Guidelines.

Honesty
We seek to promote the public good in our activities. To the best of our ability, we provide truthful and accurate communications. We dedicate ourselves to conciseness, clarity, coherence, and creativity, striving to address the needs of those who use our products. We alert our clients and employers when we believe material is ambiguous. Before using another person's work, we obtain permission. In cases where individuals are credited, we attribute authorship only to those who have made an original, substantive contribution. We do not perform work outside our job scope during hours compensated by clients or employers, except with their permission; nor do we use their facilities, equipment, or supplies without their approval. When we advertise our services, we do so truthfully.

Confidentiality
Respecting the confidentiality of our clients, employers, and professional organizations, we disclose business-sensitive information only with their consent or when legally required. We acquire releases from clients and employers before including their business-sensitive information in our portfolios or before using such material for a different client or employer or for demo purposes.

Quality
With the goal of producing high-quality work, we negotiate realistic, candid agreements on the schedule, budget, and deliverables with clients and employers in the initial project planning stage. When working on the project, we fulfill our negotiated roles in a timely, responsible manner and meet the stated expectations.

Fairness
We respect cultural variety and other aspects of diversity in our clients, employers, development teams, and audiences. We serve the business interests of our clients and employers, as long as such loyalty does not require us to violate the public good. We avoid conflicts of interest in the fulfillment of our professional responsibilities and activities. If we are aware of a conflict of interest, we disclose it to those concerned and obtain their approval before proceeding.

Professionalism
We seek candid evaluations of our professional performance from clients and employers. We also provide candid evaluations of communication products and services. We advance the technical communication profession through our integrity, standards, and performance.

FIGURE 6.4 ■ STC ethical guidelines for technical communication.
Source: Reprinted with permission from the Society for Technical Communication, Arlington, VA, U.S.A.

and 6.4. provide examples of two other ethics codes—one from the Association of Computing Machinery (ACM) and one from the Society for Technical Communication (STC).

Review Checklist

Ethics, technology, and communication	These three issues go hand in hand and are not separate from each other. How you write and design technical communication products will be based, in part, on your ethical stance.
Case examples of ethical issues in technical communication	Three-Mile Island Nuclear Power Plant Space Shuttle Challenger Pentium III Chip
Types of ethical choices	*Kant's categorical imperative*—emphasis on action, not consequence *Mill's utilitarianism*—consequence and act are both important when considering ethical behavior *Ethical relativism*—judging the ethics of a situation depends on culture and individual circumstance
Reasonable action	Basing ethics on *obligations, ideals,* and *consequences*
Legal versus ethical	Copyright issues Plagiarism Privacy
Other unethical actions	Social pressures Teamwork versus groupthink Suppressing knowledge Exaggerating claims about technology Exploiting cultural differences
Types of technical communication affected by ethical issues	Graphics Web pages and the Internet Memos Instructions Reports Proposals Oral presentations
Responding to ethical situations	Refer to reasonable criteria Refer to codes of ethics created by professionals

EXERCISES

1. Locate the professional code of ethics for your major or career. Divide into groups of three or four students, each of whom has a different major. Compare your professional codes, noting similarities and differences. Discuss why each code seems appropriate for that profession.

2. **Focus on ⟶WRITING.** Assume that you are a training manager for ABC Corporation, which is in the process of overhauling its policies on company ethics. Developing the company's official Code of Ethics will require several months of research and collaboration with attorneys, ethics consultants, editors, and others. Meanwhile, your boss has asked you to develop a brief but practical set of guidelines for ethical communication, as a quick and easy reference for all employees until the official code is finalized. Using the material in this chapter, prepare a 2-page memo explaining to your fellow employees how to avoid major ethical pitfalls in corporate communication.

3. In groups of two or three, locate a piece of technical communication (or use one provided by your instructor) and evaluate its ethical stance. Is the information presented in such a way that ideas or facts are exaggerated or suppressed? Are any cultural issues exploited? Share your thoughts in class, and explain how your team would redo the information.

The Collaboration Window

In groups of three or four, locate another piece of technical communication, but this time, actively search for one that seems to take an ethical stance in how it presents its information. If you can, locate the author or one of the writers of this piece and ask how this person made his or her decisions. If you cannot locate the author, speculate on the organizational dynamics, legal issues, and personal choices made by this person. How would members of your team go about making the same decisions?

The Global Window

Focus on ⟶WRITING. Different countries view plagiarism differently. Using the Internet, research the standards and laws of countries other than the United States with regard to what would happen if someone plagiarized. Do the same for policies about privacy and other ethical issues. Write a short report and present your findings in class.

Click on This

The following sites provide useful information as well as additional resources on ethics:

- General site on ethics: ethics.acusd.edu/index.html

- The Online Ethics Center for Engineering and Science at Case Western Reserve University: litwww.cwru.edu/affil/wwwethics

- Engineering Ethics: lowery.tamu.edu/ethics

- Professional Ethics for Scientists Annotated Bibliography: saber.towson.edu/ ~ sweeting/ethics/ethicbib.htm

Copyright and Privacy

Why technical communicators need to understand copyright and privacy

Section one: Copyright

Copyright—an overview

Documenting your sources

Electronic technologies and copyright

Section two: Privacy

Privacy: an overview

Computer technologies, privacy, and technical communication

Review checklist

The collaboration window

The global window

Click on this

Why technical communicators need to understand copyright and privacy

Technical communicators rarely create every word, image, or sound from scratch. Often, just the right diagram, image, sound, or wording will be found in some other material. You have probably had this experience when preparing a project for a class or work. You begin researching on the Web and find the perfect piece of clip art. Or you are looking through a trade magazine or newspaper, and there it is: exactly the right chart for the Implications section of your report. Or maybe you need a diagram for an upcoming presentation, and you find one in a magazine. Can you use these materials without permission? What if you scan the image into your computer and modify it first? What if the project is strictly for school: Didn't you hear somewhere that use of material for educational purposes requires no permission? But what if the project is for your company?

Along with copyright concerns, privacy is an issue for technical communicators. If you are working on a project that involves a Web site, you may face the decision of whether to collect personal information from site visitors. Can you collect this information and legally use it without the user's permission? If you are writing a manual, are you allowed to use demographic data about the organization's customers?

Many organizations have legal departments to help you answer these questions. However, if you are a student, a freelance communication consultant, or are rushing through a project without time to seek advice, you need to know the basics.

It's especially important to understand copyright and privacy in the age of electronic technologies. Although such concerns existed before the Web, they are heightened by the speed and power of these new technologies. Technical communicators are prime users of these technologies and must be aware of not only the technical aspects but also the legal ones.

Section one: copyright

Copyright—an overview

"Copyright law," says one expert, "is essentially a system of property." However, unlike physical property—your car, your home, your land—"the province of copyright is communication" (Strong, 1993, p. 1). In other words, copyright is the legal system that gives owners rights over their communication products. These products can include books, musical recordings, photographs, drawings, letters, memos, and more. Any time an idea can be fixed in a tangible medium, that communication product can be copyrighted.

Copyright law originated in England in part to protect the printing trade. The United States Congress established copyright because it seemed the best mechanism for encouraging creativity and for providing the public with a rich source of information. Copyright is essentially a system for "promoting and advancing knowledge" (Cavazos & Morin, 1994, p. 48). Authors and creators have an incentive to create new works, such as a new novel or a piece of music, because they know that for a limited time, they will own the copyright to their work, and others will be legally prohibited from copying it. After a set time, when the copyright expires, the work enters "the public domain" (more on this in the next few sections). Public domain material is accessible to everyone. This access to information is important in a democracy.

Remember that you can't copyright ideas, just expressions of ideas. As one group of copyright experts notes (Cavazos & Morin, 1994, p. 50):

> An idea or fact cannot be owned, but the unique description of the idea or fact in original terms can be. An author could not claim ownership to the idea of three pigs attempting to outsmart a big, bad wolf, but if the author writes her version of the story, [this version] becomes her property, protected by a copyright.

Naturally, the lines can get a bit blurry, and courts often determine the details of a case. But in general, the thing to remember is that you copyright the expression of an idea when the expression becomes fixed in a tangible medium. If you have an idea for a diagram of the brain, for example, and you create an original diagram (by hand or with a drawing program such as *Adobe Illustrator*), you hold the copyright to this diagram but not to the idea of the structure of the human brain or other diagrams of the brain.

Copyright differs from patent or trademark law

Copyright law is part of a broader set of laws that deal with *intellectual property*. Intellectual property is the result of creative expressions, inventions, and designs. Copyright is the arm of intellectual property law designed to deal with creative works. Two other types of intellectual property law that are not discussed at length in this book may interest you at some point in your career.

Patent law. Patent law governs mechanical inventions, machines, and processes. In many engineering and science organizations, as well as research universities, patent specialists make sure that a new invention or process is filed in the patent office.

Trademark law. Trademark law governs icons, symbols, and slogans (Strong, 1993, p. 1). Organizations file for trademark in order to protect their unique logos, such as the Coca-Cola name or symbol or the Nike "swoosh." The label ™ is used to indicate trademark.

How individuals and companies establish their copyright

First, the individual or company must create an original expression of an idea: a brochure, novel, poem, photograph, diagram. Since 1976, any communication you create is automatically copyrighted the moment the item becomes fixed in a tangible medium. In other words, the moment you type, write on paper, photograph, draw, or record something, you (or your company) automatically own the copyright.

You can help remind people of the copyright by adding this information somewhere on the product itself:

© 2001 Daniel P. Olsen

© 2001 Central Geology Corporation

To gain full legal protection, you can register the material with the Library of Congress. Should you or your company need to sue for copyright infringement, this registration provides extra evidence.

Even if an author does not add the © symbol, the work is automatically copyrighted. In other words, nearly everything you see in print or on the computer is copyrighted.

What rights a copyright holder can claim

Only the copyright holder has the right to

- reproduce the material
- create a derivative work
- distribute the work or
- display or conduct a public performance of the work

The copyright holder can give permission—limited or full—for others to use the work. So, for example, if you are interested in using the following drawing of an anticollision light power supply (Figure 7.1), you would need to contact the copyright holder, in this case, the Society for Technical Communication (STC). The drawing was reproduced from the front cover of an issue of the STC's journal. You would not, however, need permission in case of fair use (p. 109).

When you can and cannot use copyrighted material

All original works, once they are fixed in a tangible medium, are copyrighted. Any time you download, copy, scan, or otherwise reproduce an item, you may be infringing on someone's copyright. Whether or not you knew the material was copyrighted is immaterial.

You can use a copyrighted work without infringing if you obtain permission from the copyright holder. The holder might grant limited permission, to use the material one time only in one publication only, or might grant unlimited rights.

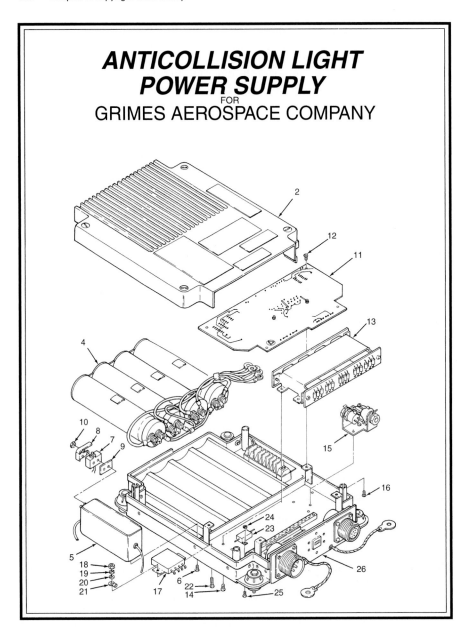

ANTICOLLISION LIGHT
POWER SUPPLY
FOR
GRIMES AEROSPACE COMPANY

Figure 7.1 ◼ **A diagram protected by copyright.** The copyright for this diagram is held by the Society for Technical Communication.

Source: Reprinted with permission from *Technical Communication,* the journal of the Society for Technical Communication.

You can also use material without permission when the materials are in the *public domain*. Copyright holders can place their materials in the public domain, or materials themselves will become public once a certain term has expired. It is common on the Web, for example, to find sites labeled "Public Domain Clip Art Files." For works created after 1977, the standard term is the lifetime of the author plus 70 years for individuals and 95 years for corporate authors (Karjala, 1999). (For works created prior to 1977, the law differs, and you can check with your legal department or a legal specialist at your school.)

Fair use doctrine. An important but sometimes overlooked legal right to use copyrighted material without seeking permission is the doctrine of *fair use*. Copyright was never intended to be a one-sided policy favoring only the rights of the copyright holders. Congress established copyright as a balance between the rights of copyright holders and the needs of the public. On the one hand, authors, artists, and other creators need incentives to produce their works. Because they hold the rights to their works for a limited time, they gain financially from their efforts. But if copyright holders were allowed full control of their materials, the public would have very limited sources of information.

So, Congress established the doctrine of fair use as a way to balance creator's rights with public access. Fair use states that under certain conditions, it is fair and legal to use copyrighted works without obtaining the copyright holder's permission. Courts ask four questions (Figure 7.2) in order to establish whether a use is fair:

- What is the purpose of the use? Commercial or educational? If commercial (and thus, for profit), courts will view it less favorably than if your use is strictly educational.
- What is the publication status of the material? If the material has been published, it will be viewed more favorably than if it has not been published. For example, your use is more likely to be considered fair if it is from a published magazine than from a series of unpublished letters.
- How much are you using? If you are using only part of a text or work, this use will be viewed more favorably than if you are reproducing the entire work.
- What will be the economic impact of your use on the original work's owner? If your use of the work will not damage the potential market of the original, this use will be viewed more favorably than if it would cause damage.

FIGURE **7.2** ■ **Guideline for determining fair use.**
Source: Based on Patry, 1985, p. vii.

Courts tend to look favorably on cases in which

- material is being used in an educational setting
- material has already been published
- only part of the material is being used
- use of the material will not affect the market value of the original

For this reason, classroom use has almost always been considered fair. Instructors and students rarely require permission to use material in a school project (but they must acknowledge the source). However, if you are a consultant or an employee of a for-profit organization, fair use doctrine may not apply.

Recently, copyright's balancing act has tipped toward the copyright holder. Fair use doctrine is often ignored. For example, if you go to the library or a copy shop to make copies, you may see an ominous warning sign above the machine. These signs rarely mention fair use. Many products, such as software, come with strictly worded statements about what will happen if you make a copy. But these statements never mention that under fair use doctrine, you have a legal right to make copies under certain conditions.

When your company owns the material

When you create technical communications as an employee, from manuals to standard operating procedures to Web pages, you are creating copyrightable material. Don't think, however, that *you* own the copyright. Under what is called the *works-for-hire doctrine*, companies in most circumstances automatically own the copyright to all materials created by their full-time employees. If you are not a full-time employee, but a consultant, the company may ask you to sign a contract stating that you will automatically give them the rights to products you produce for them. For example, if you are doing freelance writing for a nursing magazine, the publisher will probably ask you to sign your rights for that story over to the corporation that owns the magazine.

Documenting your sources

Even if you receive permission to use someone else's ideas or material, or if your use qualifies as fair use, you must document the source of this material. You can document your sources by including an in-text reference, a caption, or a statement (if you use the material during an oral presentation). For instance, if you use some copyright-free clip art in a paper for school, you don't need permission, but you should still credit the artist or company by adding a caption that includes the name of the artist or company and the location where you found the material. For more on documenting sources, see Appendix B.

Electronic technologies and copyright

Copyright is at a crossroads. On the one hand, the laws about copyright are clear: All items fixed in a tangible medium are copyrighted. You need to request permission to use these items unless they are in the public domain or meet requirements for fair use. Yet the technology trend encourages just the opposite approach. Copying, scanning, making transparencies, downloading files—all these tasks that have become so common in the workplace and on the Internet are potential violations of someone else's copyright. As a technical communicator, you will be increasingly surrounded by technology that invites you to take files, clip art, images, sounds, and more and use these in your own material.

Photocopiers and scanners

Copiers were one of the first technologies to raise serious questions about how to work with copyright in an electronic age. Most copy shops have reacted conservatively, in part due to lawsuits that accused them of violating copyright while making copies of articles and publishing these as student coursepacks. For this reason, you often see harshly worded copyright statements posted above photocopiers.

Copying something without the copyright holder's permission is often an infringement. But remember that fair use doctrine allows you to make copies for educational purposes. Even in a workplace setting, if you make some copies to distribute at a meeting, you are exercising the educational aspect of fair use. But if you wanted to make a copy of a diagram or illustration for use in your company's annual report, you would need the copyright holder's permission.

The same guidelines apply to scanners. Scanning an image is a violation of copyright unless your use qualifies as fair (see Figure 7.2) or you have permission. However, scanners and image software (such as *Adobe Photoshop*) allow you to take the scanned image and manipulate it, often to the extent that no one would ever recognize it as the original. Is this a copyright violation? It could be. But what if no one can tell? This becomes a question not just of law but also of your own ethical standards, as discussed in Chapter 6.

The Web as a marketplace of ideas and information

The Web is often billed as an information marketplace, and indeed, you can find endless images, sounds, information, and photographs on almost any topic by searching a few Web sites. The Web is part of the Internet, and the Internet has always been based on the idea of open information. In the early 1980s, when the Internet was still young, researchers and students used the technology to share ideas. The Web continues this tradition. On almost any Web site, you can click on "download source" and obtain the HTML source code for the Web page. You can also obtain the graphics, logos, diagrams, images, and more with a few simple clicks.

This technology thus encourages people to take copyrighted material without ever considering the original work's legal status. In fact, observers have speculated that the Internet might spell the death of copyright law as we know it. Instead, what seems to be happening is that copyright law is being reconsidered—and sometimes, strengthened—in light of this technology. So, even though the Web might encourage certain behaviors, it is still a good idea to keep copyright law in mind when you use material from the Internet.

Using material from the Internet

Perhaps you have located a bar chart on the Web that you'd like to use in your report. It is technically very easy to cut and paste or download the bar chart and insert it in your document. But is it legal? You notice that the chart has no copyright symbol (©). But remember: All expressions fixed in a tangible medium are copyrighted. The © symbol is not necessary. Therefore, downloading the chart and using it in your document could be a copyright infringement. You should first consider whether your use is covered by the fair use doctrine (see Figure 7.2). Is your report for commercial or educational purposes? Has the material already been published? (Web pages count as publishing.) How much of the material will you use, and how will your use affect the original? If in doubt, ask permission.

But what if, instead of using the bar chart in your printed report, you create a link to it from your company's Web page? Is providing a link to another site equivalent to copying or reproducing that site? This and similar questions related to Internet technologies are hard to answer. If you're worried about a possible lawsuit, check with an attorney who specializes in intellectual property.

Locating copyright-free clip art

You can safely use copyright-free images, sounds, graphics, or photographs, and many sites on the Internet offer such items. Although these items were copyrighted at one time, the copyright holder decided to place the materials in the public domain. The Internet encourages this kind of sharing, and you can often find exactly what you need on a copyright-free page. Usually, the author or owner will ask that you credit the source, and you should honor this request. For example, if you use an icon from such a page, you would add the line

I   Copyright-free image. Provided by www.yoursource.com

or something similar. Often, the owner will ask you to use a particular phrase.

Email and electronic messages

Remember the recent email message you received from a classmate, your boss, or your mother? That message is copyrighted by the person who created it. So, if you forwarded this message on to a friend, technically, you infringed on the owner's copyright. The same is true for a posting to a listserv, bulletin board, or other electronic discussion site.

Copyright law will probably bend at some point to accommodate this use of technology. However, if you wish to use an email message or list posting as part of a project, particularly a research or company project that will receive wide publicity, ask the copyright holder's permission (unless your use complies with the fair use doctrine shown in Figure 7.2).

The posting of email messages beyond where the original author intended also raises privacy concerns, which are addressed in the next section.

CD-ROMs and multimedia

The more media you work with, the harder the questions about copyright and intellectual property become. If your company plans to produce a CD to accompany a new product, that CD may combine images, music, film clips, and text. Each item will require careful checking to ensure that the proper legal issues have been addressed. Working with multimedia creates a labyrinth of complex legal issues (Helyar & Doudnikoff, 1994, p. 662) that requires close consultation with an intellectual property attorney.

SECTION TWO: PRIVACY

PRIVACY—AN OVERVIEW

Like copyright, privacy is a concept people may not consider until it is too late. For example, new technologies make it easier and easier to accumulate data about consumers, and until that data are misused, people rarely stop to think about how much of their personal information resides in numerous databases.

Privacy is a legal concept, and although U.S. citizens frequently talk about their right to privacy, this right is not actually stated in the U.S. Constitution. Instead, privacy law is based on court cases, state law, some federal laws, and constitutional issues such as the Fourth Amendment ban against search and seizure. One set of experts speculates that the framers of the Constitution did not explicitly write any privacy amendments because in their day, the ways in which privacy could be invaded on today's scale were impossible to imagine (Cavazos & Morin, 1994, p. 13).

Computer technologies, privacy, and technical communication

What the Constitution's framers could not foresee were the communication technologies that have been developed over the last century. As Cavazos and Morin (1994, p. 13) note,

> [e]lectronic eavesdropping devices, video and sound recording instruments, and large databases of personal information are all now a part of life in this country. Our legal system has had to react constantly to these new developments as they emerge, striving to maintain the proper balance between civil liberties and the needs of society.

Technical communicators are often at the forefront of these issues, because they are involved in designing manuals, software interfaces, Web pages that collect personal data, video training tapes, and other products that require the use of personal data or images.

Privacy in cyberspace

Web pages can raise important privacy concerns. If you are part of a technical communication team designing a company Web site, you will need to answer certain specific questions related to the privacy of those who visit your site.

Should you use cookies? Cookies are computer files that a Web site sends to the computer that has connected to the Web site (Figure 7.3). Web providers will often explain cookies in terms of their positive benefits. Cookies are helpful in that if you connect to this Web page again, the Web site already has information about your computer and thus saves you time in downloading

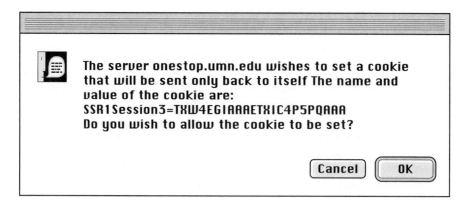

The server onestop.umn.edu wishes to set a cookie that will be sent only back to itself The name and value of the cookie are:
SSR1Session3=TXW4EGIAAAETXIC4P5PQAAA
Do you wish to allow the cookie to be set?

Cancel   OK

Figure 7.3 ■ **A cookie notification.** Web browsers can be set to warn users before they accept a cookie; however, most users don't know how to activate this feature, or even why they might want to.

Source: Reprinted with the permission of the Regents of the University of Minnesota.

information or browsing through the page. However, cookies also are problematic in that they can be used by the Web provider to obtain personal information about you and your computer—information that may not be just used by the Web owner but also loaned or sold to other companies.

When you help create a Web page, you need to decide if this page will automatically send cookies to users who connect to it. Your team or company may automatically send cookies, but you may want to think about having an overall privacy policy for the Web site instead. Many companies place such privacy statements as links on their Web sites, as in the example in Figure 7.4. Such a statement lets users know how their information will be used and what choices they have, including information on whether the site uses cookies.

As a technical communicator, you may be asked to help write such a privacy statement. Review the privacy resources at the end of this chapter and make sure you consult with the company's attorney.

Are you in compliance with global privacy laws? The Web has made it difficult for U.S. companies to ignore international issues. When you create a Web page

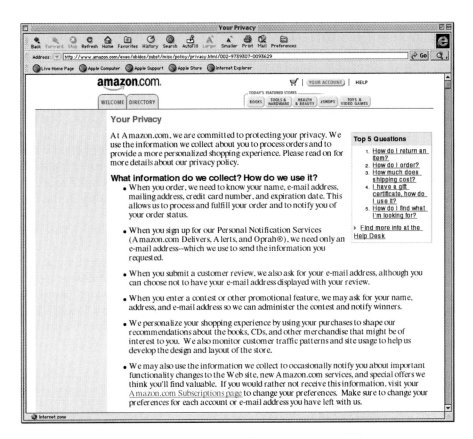

FigurE 7.4 ■ **A privacy statement from the Amazon.com Web site.**
Source: amazon.com.

and place it on the Internet, almost anyone from anywhere in the world can access the page. Despite countless benefits, this broad reach across international borders also raises concerns about privacy. Different countries and regions have different laws, viewpoints, and practices about issues that a U.S. Web developer might take for granted. Privacy is such an issue.

In the United States, collecting personal data is common practice. When you shop for groceries, the supermarket cashier will scan your special "savings card." This card not only saves you money but also allows the supermarket chain to gather information about the types and brands of items you purchase. Besides being used by the store, this information is often sold to other companies. For example, if you regularly buy a certain brand of cat food, the store may sell your name, and the names of others like you, to marketers who sell pet items.

The same scenario is true for most U.S. direct mailing services. If you make purchases through your favorite catalog, that company is allowed to sell your name and address to other companies. As most of us know, this process can generate lots of junk mail. Now, this process is in full swing on the Web. When you enter your name on a Web site as part of making a purchase, chances are that your name, address, and visits to that Web site will be used for marketing purposes and possibly sold to other companies. Cookie and other technology simplifies this sort of data collection.

In the United States, your personal information is up for grabs unless you specify that you do not want it sold. You can file your name with a special service, and they will attempt to take your name off most mailing lists (but the process is never perfect). And, under certain laws, especially when your credit history is involved, you have a legal right to correct misinformation. But the burden is on you to take action.

In Europe, however, companies are not allowed to collect information without your permission. They must also provide a clear mechanism for you to change or remove your information. The European Union's Data Privacy Act turned this policy into law. Anyone doing business with a country in the European Union must follow this law.

So, when a citizen of the European Union logs onto a U.S.-based Web site, and when that Web site sends cookies or collects personal information, the company sponsoring the site may be violating the European citizen's legal rights.

In short, going by the laws and habits of U.S. citizens is not always appropriate in designing Web pages with a global reach. To accommodate non-U.S. visitors to your Web site, work with the company's attorney or with colleagues in marketing or international sales. Consider formulating a specific privacy policy for your company Web sites.

Privacy and documentation

Technical communication and privacy also intersect in documentation. When you prepare documentation (manuals, quick reference cards, brochures, help screens), you may wish to include quotes from customers or examples of how the product is used in other companies. But before you can use such information, you need written permission from the individual customers and/or the companies involved. The same is true if you use any photographs of customers using your product. While you or your company may hold the copyright to such items, you still need to ask permission so that you are not violating the individual's or company's right to privacy.

Privacy and videotapes

Technical communicators are often involved in helping to prepare training tapes. These tapes are used by company trainers, marketing and sales staff, or customers to help people learn to use a new product or service. For example, if you work for a company that builds gardening equipment, your company may create both a manual and a training tape to accompany a rototiller. Customers could read the manual or watch the tape to learn how to operate and maintain their new machine.

As part of the training tape, you may wish to use video footage taken when your marketing team was out visiting customers. Or you may have other video footage of employees in the company using the equipment. Even though your company may own the copyright to this material, it is still important to obtain permission from the people in the video (employees or customers) to use the footage in a training video.

Review Checklist

Copyright	Assigns rights to reproduce or distribute material to a person or organization; differs from patent and trademark law; covers any "expression of an idea fixed in a tangible medium." Documentation, photographs, reports, email messages are all copyrightable.
	Is automatically established when the item is created. Registering with the Library of Congress provides additional protection.
When can you use copyrighted work?	When you have the owner's permission.
	When your use is governed by the *fair use doctrine* according to these criteria: • Purpose (commercial or <u>educational</u>) • Publication status (<u>published</u> or unpublished) • Amount you will use • How your use will affect the value of the original

(continued)

When does a company own the work?	If you are a full-time employee, companies generally own your work under the *works-for-hire doctrine.*
How do electronic technologies complicate copyright?	Photocopiers, scanners, material from the Internet—all create the potential for copyright violations. Look for *public domain* material if you can locate it.
Privacy	Privacy is often overlooked until problems develop. Technical communicators need to be aware of the privacy implications of the products they work on.
How do electronic technologies complicate privacy issues?	Web pages, cookies, data collection—all potentially undermine individual privacy.
How are documentation and videotapes affected by privacy issues?	Make sure you obtain permission to use an image of a person in documentation or videos, even if your use is legal under copyright law.
What are the international implications of electronic data collection?	The European Union has a very different perspective and very different laws on the use of personal data.

 ## Exercises

1. Interview a professional in your field to determine how copyright law affects his or her work and company. Consult with the company's legal department if possible. Report your findings in a memo to your instructor.

2. In groups of two or three, discuss two possible situations in which someone else's material might be used for a new purpose—for example, a diagram from a magazine could be scanned in and used for a brochure. Given the details of the situations you describe, decide if the material is being used according to fair use guidelines.

3. Focus on ⮕ Writing. Use a Web search engine to locate sites that describe copyright disputes. Discuss these in class, and determine if any of the disputes are relevant to technical communication. Design a brief handout or a simple Web page describing one major dispute and turn this in to your instructor.

 ## The Collaboration Window

Form groups according to your major and plan a brochure (or some other form of technical communication) using copyrighted materials. Write up a process by which you would obtain the materials (public domain, request permission). Share your findings with the rest of the class.

The Global Window

Copyright and privacy are subject to the laws of a country or region. The Internet and other technologies complicate matters because data can easily travel around the world without regard for borders. The World Intellectual Property Organization (WIPO) is an international group that considers the global implications of intellectual property issues. You can learn more about WIPO by visiting their Web site, located at www.wipo.org.

If you can, talk to a communication professional who deals with international audiences. Ask about any difficulties she or he has ever encountered with copyright or privacy. Review the WIPO Web site with this professional, and ask how the information on this site might affect the way she or he would write, design, and distribute a document.

Click on This

Use your favorite search engine to locate sources of graphics, sounds, and photographs on the Web. Examine the sources and determine which offer copyright-free images and which do not. Compile a list of your findings and post this list to your class Web site (or create a handout to distribute in class). A few sites to help get you started are:

www.imageclub.com/clipart/catalogs/index.html
www.peisland.com/fyi/clipart/clipart.html

Also, the following sites provide useful information on copyright and privacy issues:

www.cpsr.org
www.epic.org
www.ncte.org/cccc/cccc-ip
www.gseis.ucla.edu/iclp/hp.html
www.loc.gov/copyright

CHAPTER **8**

Page Layout and Document Design

CREATING VISUALLY EFFECTIVE DOCUMENTS

TYPOGRAPHY

PAGE LAYOUT

CREATING AN EFFECTIVE TABLE OF CONTENTS AND INDEX

CREATING AND USING STYLE SHEETS

ORGANIZATIONAL STYLE GUIDES

USING WORD-PROCESSING AND PAGE LAYOUT SOFTWARE

DESIGNING ELECTRONIC DOCUMENTS

REVIEW CHECKLIST

EXERCISES

THE COLLABORATION WINDOW

THE GLOBAL WINDOW

CLICK ON THIS

Creating visually effective documents

As with all decisions related to technical communication, page layout and document design are based on informed choices. You make decisions about fonts, format, headings, page size, and the like based on the document's audience and purpose. For example, imagine that you are asked to produce a brochure for a physician's office. The brochure's audience is patients who have heart conditions and may need heart surgery. The brochure's purpose is to explain the procedure, answer frequently asked questions, and help explain the risks while reassuring patients.

Before you would even begin to think about what to write or how to format the document, you would do an audience and purpose analysis (Chapter 2). If you learned that the brochure also needed to include a list of tasks, such as a checklist of procedures patients need to perform, you might also conduct a task analysis (Chapter 3). Your analysis would yield important information. For example, you might learn that

- patients are frightened
- patients are often older and may have difficulty reading small type

These items would be important not only as you wrote the copy, but also as you thought about designing the document. Frightened patients might prefer to read a document that is soothing to look at. You could choose a comforting typeface, pleasant graphics, and a warm color for the paper. Also, because your audience has trouble reading small type, you would make sure the font is large enough to be read easily.

As you can see, document design, like all technical communication, puts audience and purpose first. In designing your document, consider the features discussed below.

Typography

The style of type you choose makes a big difference in how audiences read and react to your document. Typefaces have personalities. Some convey seriousness; others convey humor; still others convey a technical quality. Before desktop computing, the art of typesetting was in the hands of skilled graphic artists and typographers who were trained in selecting and using fonts for the most effective results.

Since the advent of the personal computer and, more importantly, desktop laser and inkjet printers, almost anyone can create documents using an array of typefaces. Most people choose a typeface based on intuition, or they mimic something seen in print (such as a newspaper or a letterhead). But technical communicators need to choose typefaces that are appropriate for the document's audience and purpose.

Fonts and families

Before you learn anything else about using type, you should be aware of the terminology. *Typeface* refers to an entire family, such as Times Roman or Helvetica. *Font* refers to one specific font within that family, such as Times Roman 12-point italic or Helvetica 14-point bold. In general, there are three categories of typefaces: serif, sans serif, and display. Each category has a specific history and is appropriate for certain audiences and purposes.

Serif type. Serif refers to the small "feet" or horizontal strokes that extend from the main strokes of a letter.

Serif type is the oldest variety. One set of experts notes that "[a]lthough the origin of serifs remains a mystery, stone carvers have traditionally used serifs to square off the ends of letters with their chisels" (Kostelnick & Roberts, 1998, p. 126). Others suggest that early fonts were designed to look very much like the handwritten calligraphy used by scribes to create the books that preceded the printing press: Readers were more willing to accept a typeface with which they were already familiar (Steinberg, 1996, pp. 9–10). Most experts agree that today's serif fonts reflect designs created in the 15th century, after the invention of the printing press.

Examples of serif typefaces include

- Times
- Garamond
- Goudy
- Baskerville
- Palatino

Figure 8.1 is an example of several available fonts in the Times typeface. "Roman" refers to the standard, nonbolded, nonitalicized font.

When to use serif typefaces. Serif typefaces, because they are the oldest, convey a *sense of formality*. They are commonly used on formal letterheads or in wedding or graduation invitations. Serif fonts are also considered somewhat more *readable*, especially for Western audiences who are accustomed to this font, because the serifs guide the eye from letter to letter. Thus, serif type is also an effective choice for *body text,* such as the main text in a report. Most newspapers use serif type, usually Times Roman, for body copy.

ABCDEFGHIJKLMNO (Roman in caps)

abcdefghijklmnop (Roman in lowercase)

abcdefghijklmnop (Italic)

abcdefghijklmnop (Bold)

FiＱuRE 8.1 ■ Times is an example of a serif typeface.

Sans serif type. *Sans* is the French word for "without." Sans serif typefaces lack the "feet," or horizontal strokes, of a serif typeface.

Sans serif typefaces were designed later than serif fonts. They became popular during the 20th century, a time when science and technology were on the rise. Their lack of the traditional serif reflected a change in mood, a more modern feeling for a more modern time.

Examples of sans serif typefaces include

- Helvetica
- Frutiger
- Stone Sans
- Gill Sans

Figure 8.2 is an example of several available fonts in the Helvetica typeface.

ABCDEFGHIJKLMNO (Roman caps)

abcdefghijklmnop (Roman lower case)

abcdefghijklmnop (Italic)

abcdefghijklmnop (Bold)

FiＱuRE 8.2 ■ Helvetica is an example of a sans serif typeface.

In the United States, sans serif typefaces convey a sense of *modernism*. They also convey a *technical* feeling to your document. Sans serif type can be used for body copy, but only in small quantities (in a brochure or short memo versus in a longer document such as a report or newspaper). Sans serif fonts are excellent for *captions*, *figures*, *labels*, and *marginal comments*. Sans serif fonts also work well for electronic documents, such as Web pages or help screens, because they are usually easier to read on computer screens with poorer resolutions.

Display typefaces. Other modern typefaces fall somewhere in between serif and sans serif, and these are often called *display*. One example is a typeface called Optima. While some might consider Optima sans serif, closer inspection reveals characters with slight serifs, much less obvious than the full horizontal strokes of the traditional serif fonts.

M N O P Q R —— NO TRUE SERIFS,
BUT SLIGHT FLARE

Display typefaces are often used for advertising or marketing materials, in order to give a special effect or feeling. Typefaces such as Optima are modern yet less harsh than a complete sans serif font, such as Helvetica. Thus, Optima or other display typefaces can give a modern look while being a bit easier to read in larger passages of text.

There are a wide variety of display typefaces, in everything from fonts that resemble handwriting to fonts that seem to move playfully on the page. Figure 8.3 samples some of these typefaces. In general, do not use these for body type, or your audience will have trouble reading. You can use them as headings, but only sparingly, or the excessive visual motion will overwhelm your audience.

THIS IS A TYPEFACE

This is a typeface

This is a typeface

This is a typeface

FIGURE 8.3 ■ **A sampling of decorative typefaces.** You can choose from thousands of typefaces and fonts using any of several CDs or desktop publishing programs.

A typeface sends a message

More than simply creating words on paper or on the screen, typefaces send messages. Some, such as serif fonts, are formal. Others, such as the decorative typefaces shown in Figure 8.3, are playful and powerful ways to draw an audience's attention to a section of the page.

As you choose your typeface, consider the document's purpose. If the purpose is to help patients relax, use a combination that conveys ease. Fonts that imitate handwriting are often a good choice although they can be hard to read if used in lengthy passages. If the purpose is to help engineers find technical data quickly in a table or chart, use Helvetica or some other sans serif typeface—not only because numbers in sans serif type are easy to see, but also because engineers will be more comfortable with fonts that look precise.

Certain fonts go together

With all these typefaces to choose from, many computer users go wild and mix fonts from unrelated typeface families. When you do this, what you end up with is a document that looks like alphabet soup: a jumble of letter forms, sizes, and shapes. Too many fonts from too many type families create visual noise.

Follow these basic rules when mixing and matching typefaces within a document:

- **Use fonts from only one typeface.** The safest rule is to stick with just one typeface. For example, you might decide on Times for an audience of financial planners, lawyers, or others who expect a traditional font. In this case, use Times 14-point bold for the headings, 12-point regular (Roman) for the body copy, and 12-point italic for the titles of books or, sparingly, for emphasis.
- **If you mix different typefaces, be consistent.** If the document contains illustrations, charts, or numbers, use Helvetica 10 point for these. Helvetica is good for captions and numbers.

  You can also use one typeface for headings and another for text. A common approach is to use Helvetica or another sans serif typeface for headings and Times Roman or another serif font for body copy, as shown in Figure 8.4.
- **Use italic, bold, and ALL CAPS sparingly.** Various types of emphasis call for italic, bold, or capital letters. For example, italic is used to set the titles of books apart from the rest of the text, as in:

  I   We read *The Grapes of Wrath* in English class.

Sometimes, people use italic much as they would use a highlighter or magic marker, to set words apart and draw attention to these words. Using italic too frequently can make it difficult for your audience to decide what information is truly important.

Implications for new technologies

Our research showed ther alkdja lajdfklasdjf kjafd jdsf jaldf akljdf alkd andoienoije hahd and and adhjnis sio bahdh a dno anf afdf afj ldjf

Budget

We have a significant increase in budget this year. For example, andka akdht is thia ud ai and a fa dfajf jafdk j;af and shs ai si ai djafoiu eiua

Timeline

We plan to implement the akld al dna kjd kajfd by ad fjafd aie ainmoieuoiua ou dl af ldjfankja l g kjd

Figure 8.4 ■ **Different fonts for different elements.** This sample page from a report uses a sans serif font for headings and a serif font for body copy.

The same is true for **bold face.** There are fewer rules about when to use bold, but too much bold will lose its appeal. Bold is good for headings, subheadings, or parts of a sentence you want to emphasize. Within text, use bold selectively.

Avoid all capital letters for more than a few words or when creating short headings, because long strings of uppercase letters and words have the same visual outline (Felker, Pickering, Charrow, Holland, & Redish, 1981, p. 87) and thus make the material difficult for an audience to read.

A short line of all capital letters may be fine, as in

❚ MY DOG IS A COCKER SPANIEL.

But the longer the passage, the harder readers must work to understand the information. For example,

❚ ACCORDING TO THE NATIONAL COUNCIL ON RADIATION PROTECTION, YOUR MAXIMUM ALLOWABLE DOSE OF LOW-LEVEL RADIATION IS 500 MILLIREMS PER YEAR.

This passage loses its impact, whereas in the next passage, a selective use of capital letters (used for the proper name of the Council and to highlight the time frame) is more effective:

According to the National Council on Radiation Protection, your maximum allowable dose of low-level radiation is 500 millirems PER YEAR.

Keep in mind that on the Internet, using all caps is considered shouting, so only use capital letters when you want to selectively create emphasis.

Choosing a type size for readability

Traditionally, type is measured in a system of *picas* and *points*. These measurements come from the days when printers used lead type and special rulers to measure how many words would fit on a page. Point sizes are familiar to anyone who uses word-processing software. Point sizes are also important if you need to mark up your copy for a compositor or printer.

Generally, the point system is used to measure a typeface. So, a 12-point font is smaller than a 24-point font. This system of measure indicates the approximate distance between the *ascender* and the *descender* of a letter form, as in:

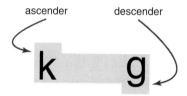

One point is equal to 1/12 of a pica and and one pica is equal to about 1/6 of an inch.

The following readability guidelines are appropriate for *most* printed documents:

- Do not go below 12-point type for most body text. (You can use 10-point, but 12 is better.) In some typefaces, 12 point will be a bit small, so base your decision on how the font looks when printed.
- Make headings at least 2 points larger than body text.
- Double space between lines of text.
- Use italic, bold, and capital letters sparingly.

For overhead transparencies or computer displays in oral presentations, consider even larger sizes: 18- or 20-point type for body text, and 20 or greater for headings. For choosing type for computer displays, including Web pages, see "Designing Electronic Documents" later in this chapter.

Using customary formats for specific purposes

In some situations, you will have little choice about the typeface you use because you may be constrained by specific conventions for that audience, document type, or situation. For example, newspapers usually follow standard conventions: articles use the same fonts for all body copy and often use a limited set of fonts for headlines. Newspapers use standard column lengths for their stories. Organizations often have standard formats for their publications and specific formats for documents such as user manuals, memos, standard operating procedures (SOPs), and descriptions (such as job descriptions). These corporate standards for type and layout are often described in company *style guides,* discussed later in this chapter.

Page Layout

The term "page" is a fluid one when you consider electronic documents: What is a page? On the computer screen, a page can go on forever. Also, *page* might mean a page of a report, but it can also mean one panel of a brochure or part of a quick reference card. The following discussion focuses primarily on traditional paper (printed) pages. See "Designing Electronic Documents" later in this chapter for a discussion of pages in electronic documents.

The design and layout of a page play an important role in how your audience will react to and interact with the information. If a page is organized poorly, readers won't be able to find what they need. And if a page or document is unattractive, readers won't be enticed to look at it. Readers expect each page to be visually appealing, logically organized, and easy to navigate.

Page and document design is an art that takes practice. More and more software is being developed to help people with page layout and document design, but all the software in the world won't help unless you know something about how readers deal with text on a page.

How readers view a page

When you design information, consider how most readers look at a page. Generally, they look at it first as a whole unit, scanning the page quickly to get a sense of the overall layout, look, and structure. Readers try to make sense of the document and determine its "road map." They ask questions such as these:

- *What are the primary headings?*
- *Where are the tables and charts?*
- *What is the main title?*

Visual hierarchy. By asking themselves these questions, your audience is trying to determine the *visual hierarchy* of the page based on its layout and design. The main items of importance should be in the primary headings, and the secondary items in the second level of headings.

As you create your document, determine which subject areas constitute the main (first), second, or third level of headings. (More than three levels of headings can confuse your audience.) Make sure that headings at the same level use the same font and are indented.

Look back at Figure 8.4 for an example of a page that provides a clear visual hierarchy. Even if you view it from far enough away that you can't make out the actual words, you can still see that the page has a structure: main headings and body text.

Gutenberg diagram. In Western cultures, people read from left to right. The eye enters the page at the upper left corner, and exits at the lower right corner, as shown in Figure 8.5. This pattern is generally called the Gutenberg diagram after Johannes Gutenberg, the inventor of movable type.

Using a grid structure

Grids help readers make sense of material, because they create an underlying structure for the page. For instance, a two-column newsletter would use a grid like the one shown in Figure 8.6.

Creating areas of emphasis

Your audience will read a page based on the Gutenberg diagram (Figure 8.5). But even though the eye enters the page in the upper left corner, it is quickly drawn to areas of emphasis: large type, color areas, boxed items (Figure 8.7), and areas surrounded by white space. If you want your audience to pay particular attention to one section of the page, such as a warning or safety item, set this material in a box or use larger type.

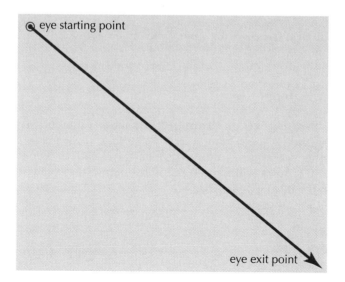

FIGURE 8.5 ■ **The Gutenberg diagram represents the way Western readers skim a document.**

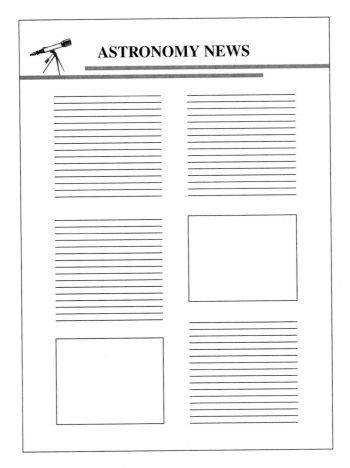

ASTRONOMY NEWS

FIGURE 8.6 ■ Grid structure for a two-column newsletter.

Using white space

Sometimes, it's what's *not* on the page that makes a difference. Areas of text surrounded by white space draw the reader's eye to those areas, partially because the white space breaks up the regular visual pattern. For example, the letter in Figure 8.7 uses white space and a box to draw the eye to important information in the middle of the page.

Using lists

Whenever you find yourself writing a series of items within a paragraph, consider using a bulleted or numbered list instead, especially if you are describing a series of tasks or trying to make certain items easy to locate. Don't overuse bullets, however, and don't use fancy icons when a plain round dot will do.

UNIVERSITY OF MINNESOTA

Dear Employee,

As you probably know, your current parking lot, SC-158, will be closed from June 1999–June 2000 to allow us to construct the new three-tier parking ramp. When the ramp is completed, you will again be able to park in SC-158. Until that time, you have been reassigned to parking lot SC-150.

> **Please note that when parking in your new location (SC-150), you will need to use the enclosed parking hang tag.**

If you have any questions, please call Parking Services at 641-2110.

Thank you.

Fɪɢᴜʀᴇ **8.7** ■ **This letter uses a box to make effective use of white space.**

Most audiences prefer that bulleted lists have a streamlined look. Figure 8.8 compares in-text lists to bulleted and numbered lists.

There is no consensus on how to punctuate a list. Style guides, which writers and editors use to check on punctuation and grammar issues, often disagree on this point. Many companies use their own style guides (see below) and include directions for punctuating a list. In general, the move is away from any punctuation within the list itself. But check with an editor in your organization or consult the company style guide.

Using headings

Readers of a long document often look back or jump ahead to sections that interest them most. Headings announce how a document is organized, point readers to what they need, and divide the document into accessible blocks or "chunks." An informative heading can help a reader decide whether a section is worth reading (Felker, Pickering, Charrow, Holland, & Redish, 1981, p. 17). Besides cutting down on reading and retrieval time, headings help readers remember information (Hartley, 1985, p. 15).

Text	Bulleted list
The fire ant has now spread to three regions: the Galapagos islands, the South Pacific, and Africa.	The fire ant has now spread to three regions: • the Galapagos islands • the South Pacific • Africa

Text	Numbered list
There are three steps to installing your modem: Open the computer case, insert the modem in the slot, and close the case.	Installing your modem 1. Open the computer case 2. Insert the modem in the slot 3. Close the case

Figure 8.8 ■ **Lists.** Use a bulleted list for three or more items in a series. In cases where you are instructing an audience to perform a series of steps, a numbered list is more appropriate.

Size headings by level. Like a good road map, your headings should clearly announce the large and small segments in your document. When you write your material, think of it in chunks and subchunks. When you analyze the document's purpose and your user's intended tasks, you will generally create an outline of your document. An outline of a report for physicians on new medications for depression might begin as follows:

Background: Current medications and their history

Recent research into new medications

Ongoing research and medications on the horizon

These are your primary, or level one, headings. Your document might also contain subheadings for each section. You can use the marks h1, h2, and h3 to indicate heading levels.

h1. Current medications and their history
  h2. Medications before the 1980s
  h2. Selective Seritonin reuptake inhibitors (SSRIs)
    h3. Prozac
    h3. Effexor
h1. Recent research into new medications
  h2. Refining the SSRI approach
  h2. Research on brain chemistry

h1. Ongoing research and medications on the horizon
    h2. Future trends in treatment of depression
        h3. Medical
        h3. Psychological
    h2. Research implications

You would then design your document so that the heading levels are consistent. All h1 headings would use the same font and indent, as would all h2 and h3 headings. Your final document might look something like this:

CURRENT MEDICATIONS AND THEIR HISTORY

Currently, several medications are popular andk dfkja fdkjdf jdasfl dsfl ls ldf jdsfjdfsjssdfkjakd aoiieu joieu ajoidu oi eruyao ghoikh ahogy henkajkd kanjkdntheoi heiou iao9j8ajfl d ald oaj glnkeyhioj

Medications Before the 1980s

Several medications were used prior to the 1980s, when fkja sdfkjakd aoiieu joieu ajoidu oi eruyao ghoikh ahogy henkajkd kanjkdntheoi heiou iao9j8ajfl d ald oaj glnkeyhioj idu oi eruyao ghoikh ahogy

Selective Seritonin Reuptake Inhibitors (SSRIs)

The increased research around the importance of seritonin fkja kd aoiieu joieu ajoidu oi eruyao ghoikh ahogy henkajkd kanjkdntheoi heiou iao9j8ajfl d ald oaj glnkeyhioj idu oi eruya

Prozac—The increased research around the importance of seritonin fkja kd aoiieu joieu ajoidu oi eruyao ghoikh ahogy henkajkd kanjkdntheoi heiou iao9j8ajfl d ald oaj glnkeyhioj idu oi eruyao gho

Effexor—The increased research around the importance of seritonin fkja kd aoiieu joieu ajoidu oi eruyao ghoikh ahofakljl gy henkajkd kanjkdntheoi heiou iao9j8ajfl d ald oaj glnkeyhioj idu oi eruyao gho

RECENT RESEARCH INTO NEW MEDICATIONS

New research indicates that andk jdf jdasfl dsfl ls ldf jdsfjdfsjssdfkjakd aoiieu joieu ajoidu oi eruyao ghoikh ahogy henkajkd kanjkdntheoi heiou iao9j8ajfl d ald oaj glnkeyhioj

Refining the SSRI Approach

The increased research around the importance of seritonin fkja kd aoiieu joieu ajoidu oi eruyao ghoikh ahogy henkajkd kanjkdntheoi heiou iao9j8ajfl d ald oaj glnkeyhioj idu oi eruya

Choose an appropriate heading size. Headings, especially at the first and second level, should be larger than the body copy they accompany. A 2-point spread is the generally accepted rule: If body copy is 12 point, headings generally should be at least 14 point. As with all decisions, base this choice on audience and purpose. Some companies, for example, might determine that for their customers a different arrangement is appropriate.

Address reader questions. Chapter 3 discusses the technique of creating headings in the form of reader questions. This approach may not be appropriate for all documents, and overuse of the technique can sound repetitious and become annoying to readers. But for some documents, questions can help guide the reader to the appropriate section of the document. For example, a patient information brochure about an outpatient surgical procedure called a laparoscopy might use question-style headings such as

WHAT ACTIVITIES CAN I PERFORM AFTER MY LAPAROSCOPY?

These headings create a more user-friendly document. Whenever you have a situation in which a user will approach a document with a series of questions in mind, consider using question-style headings.

Create headings as points of access to the text. Headings point readers to the information they need. Readers often flip through the pages of a document and look at the headings to determine what section they wish to look at.

Headings can be placed in the left of a page, creating what is called the cueing area—a place where readers stop, examine the topic, and then begin reading the text at that location. Figure 8.9 shows a page that uses a left-margin cueing area. The blank space around the heading is also useful as a place where readers can make notes to themselves or where you can place an icon as a way of drawing reader attention to this area.

Some documents use running heads or feet—headings that not only appear in the text but also across the top or bottom of each page (see Figure 8.10). Readers of technical documents often open the manual, handbook, or report in the middle of the document, skimming through pages and looking for what they need. Running heads give these readers a clear sense of what chapter and section they are in.

Make headings visually consistent and grammatically parallel. Heading levels should be consistent. For example, level one headings are 12 point, bold upper case and set flush left with the margin; level two headings are 12 point, bold in upper and lower case indented one tab setting; level three headings are 10 point bold, indented, with the text run in.

Along with visual consistency, headings of the same level should also be grammatically parallel (see Chapter 3). For example, if you phrase headings in the form of reader questions, make sure all are phrased in this way.

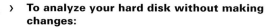

5-4 *Norton Utilities for Macintosh*

Analyzing your hard disk

› **To analyze your hard disk without making changes:**

1 **Start Speed Disk by double-clicking the Speed Disk icon in the Norton Utilities 2.0 folder.**
   If you're already in Norton Utilities, and working from your hard drive, choose Speed Disk from the Utilities menu. The Speed Disk window appears.

   Do not try switching to Speed Disk if you are using the Emergency Disk; the Emergency Disk contains only the Norton Utilities.

   Left-margin cueing area

   Drag the icon of the disk you wish to optimize onto the Speed Disk icon. Speed Disk starts analyzing the chosen disk.

2 **Click the Check Drive button.**

   Check Drive

   Speed Disk starts analyzing your disk. The progress bar indicates Speed Disk's activity as it examines files. When it is finished, you see a comment in the message area on the right.

3 **Note Speed Disk's comments in the message area on the right.**
   Speed Disk reports the amount of fragmentation and suggests whether to optimize.

   | Largest Free Contiguous | 6.5 M |
   | File Fragmentation | 0.07% |

   To take a closer look at the organization of files on your hard disk, choose Go To Expert from Speed Disk's Options menu. See the Reference section for more details.

4 **To quit Speed Disk, choose Quit from the File menu.**
   You quit Speed Disk and the Norton Utilities.

   You can also choose another utility from the Utilities menu. If you want to continue and defragment files or optimize your hard disk, read on.

FiGURE 8.9 ■ Left-margin cueing area.
Source: Symantec Corp.

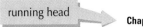running head   **Chapter 12**   Glossaries: Storing Items for Reuse   **259**

To open a glossary file

1   From the Edit menu, choose Glossary.

2   From the File menu, choose New to clear all entries except those supplied with Word.

3   When Word asks if you want to delete all nonstandard entries, choose Yes. Deleting the nonsupplied entries does not remove them from the original, opened glossary. It only clears them so that they are not merged into the glossary you open.

4   From the File menu, choose Open.

5   From the list in the Open box, select the glossary you want.

To open a glossary and merge it with the current glossary

1   From the Edit menu, choose Glossary.

2   From the File menu, choose Open.

3   From the list in the Open box, select the glossary you want.

When you merge a glossary with the current glossary and both glossaries have an entry with the same name, Word uses the text from the second glossary for the glossary entry.

To start a new glossary file

1   From the Edit menu, choose Glossary.

2   From the File menu, choose New to clear all entries except those supplied with Word.

3   Create as many glossary entries as you like.

4   When you're ready to save the new glossary, choose Glossary from the Edit menu.

5   From the File menu, choose Save As.

6   Type a name for the glossary in the Save Glossary As box.

7   Choose the Save button.

Microsoft Word User's Guide

Figure **8.10** ■ **An example of a running head from a word-processing manual.**
Source: Microsoft Corp.

CREATING AN Effective TAble of CONTENTS ANd iNdEX

Part of the page layout and design process often involves creating a table of contents and an index, especially for large reports, books, or similar publications. Even more than headings, tables of contents and indexes become the primary access points by which a reader will enter your text. Generally, readers start by checking the index. Next, they try the table of contents. If they still cannot locate the information, they begin paging through the material at random, hoping to find something.

Most word-processing and page layout programs can automatically generate a table of contents or an index. You insert markers (tags) in the document, and the software will look for the information in these tags and compile it into a table of contents or an index. The key to a successful table of contents and index, however, is not in the computer software but in knowing what categories, words, and topics your audience will search for when using these tools. Most of us have had the frustrating experience of trying to look something up in an index and discovering that the term we use is not the term the index uses.

It is best to have your table of contents reviewed by an editor (for problems with parallelism or gaps in coverage) and to have your index prepared by a professional indexer. Many companies have technical writers on staff who specialize in these areas. You can also hire freelance editors and indexers to do these jobs.

CREATING ANd usiNG style sHEETS

Style sheets are helpful guides that technical communicators use to ensure consistency across a single document or a set of documents. It's not as hard to maintain consistency if you are the only writer, but if you are working as part of a team (usually the case), it's important to be sure that all the writers are using the same typefaces and fonts in the same manner. A style sheet can be as simple as a one-page handout, similar to Figure 8.11. The more complex the document, the more specific the style sheet should be. All writers and editors should have a copy of the style sheet. Consider keeping the style sheet on a Web page for easy access and efficient updating.

ORGANIZATIONAL style GuidES

In addition to style sheets for specific documents, some organizations produce style guides containing rules for proper use of trade names, appropriate punctuation, preferred fonts and typefaces, and so on. Style guides help ensure a con-

Style sheet for the EcoSystems water filter manual

Body text

Body text is set in 12 point Palatino, flush left, ragged right.

Level one heading

Palatino 12 point, Roman, first initial cap, flush left. Body text follows on the next line with one tab indent.

(sample)

How to install the water filtration sytem

Bland alkjdf a akdka diosa sisi sht ek adie u afj;aiudnafdjaf If a snetne cekt ois ios fja;jdfkjafdd.

Level two heading

Palatino 12 point, Roman, underlined, first initial cap, one tab indent. Body text follows on the next line with one tab indent.

(sample)

Perform an audience and purpose analysis

Bland alkjdf a akdka diosa sisi sht ek adie u afj;aiudnafdjaf If a snetne cekt ois ios fja;jdfkjafdd.

Fiɢᴜʀᴇ 8.11 ■ Simple style sheet.

sistent look across a company's various communication products. Style guides need to be updated regularly to reflect the new products and services an organization might offer. Sometimes, a technical communicator will learn one method of using typefaces or designing pages only to change jobs and discover that the new company follows a different set of rules. In the end, you must go with whatever the company prefers. See Figure 8.12 for a sample style guide page.

Using word-processing and page layout software

You can choose from a wide range of software when designing and laying out a document. Most technical communicators use either word-processing or page layout software, although many use markup languages.

Word-processing and page layout software

Word-processing software, such as *Corel WordPerfect* or *Microsoft Word* can handle more than words. These programs allow you to incorporate graphics, tables, images, fonts, and color. For larger projects, such as books with many chapters, or for projects that involve complicated types of layout, many people choose page layout (or desktop publishing) software such as *PageMaker, Adobe Framemaker,* or *Quark.* Technical writers on large projects prefer this software because it allows for more sophisticated use of certain features such as templates (formatted styles and page designs that can be shared among different documents). Page layout software is harder to learn to use than word-processing software, but it can be a far better tool for managing large or complex projects.

Markup languages

For projects that will be shared across different types of computer platforms, many technical writers prefer markup languages. These languages use marks, or "tags," to indicate where the text should be bold, indented, italicized, and so on. Word-processing or page layout files from different programs or platforms are not always compatible, but with markup languages, once the tags are inserted, documents can be shared across many platforms. Also, using tags allows for more flexibility in how a document looks. For instance, the following tag might be used to indicate that a line of text is a level one heading:

```
<h1>Installing the replacement engine</h1>
```

5 Grammar and Punctuation

41 Introduction
41 General Guidelines
47 Punctuating Words and Sentences
57 Capitalizing Words
59 Hyphenating Words
62 Dividing Words

Introduction

This chapter summarizes general rules of grammar and conventions of correct writing. Issues such as number agreement, parallel construction, proper punctuation, and proper division of words are covered here. Notice that in some cases examples appear in a *Wrong* versus *Right* format. Although those examples labeled *Wrong* are sometimes grammatically or technically correct, they violate the rules or conventions of SAS Institute and should be avoided.

General Guidelines

Articles The following rules govern the use of articles:

- Be careful to use the appropriate article to modify nouns in your writing. Articles can be considered adjectives because they modify the nouns they designate. Indefinite articles (*a* or *an*) do not limit the nouns they modify because they do not specify a particular case. The definite article (*the*) limits the nouns it modifies by making them more exact.

    You must sort *a* data set before using the BY statement.
    (In this sentence, you are not sorting a specific data set.)

    You must sort *the* data set before using the BY statement.
    (In this sentence, you are sorting a particular data set.)

- If you use a series of adjectives to modify a noun that represents more than one item or more than one group of items, repeat the article before each adjective.

    We bought *the* translated and *the* illustrated handbooks.
    (We bought two different kinds of books, some that are translated and some that are illustrated.)

Figure 8.12 ■ **Sample pages from a style guide.**
Source: Reprinted with permission: SAS Institute Inc., *Corporate Identity: Written style,* Second Edition, Cary, NC: SAS Institute Inc., 1989, 212 pp.

However, if you use a series of adjectives to modify a noun that represents only one item or one group of items, use one article before the series of adjectives.

> We bought *the* translated and illustrated handbooks. (We bought some books, all of which are both illustrated and translated).

Number Agreement Number agreement is an important aspect of effective writing. Use the following rules to produce agreement in your writing:

□ Make subjects and verbs in sentences agree grammatically. Singular subjects take singular verbs; plural subjects take plural verbs.

> Singular: That *observation is* not significant to the test.
>
> Plural: Those *observations are* not significant to the test.

Note: Many organizations use the word *data* with a singular verb. At SAS Institute, *data* always requires a plural verb.

> Wrong: The *data is* stored on disk or tape on the main computer.
>
> Right: The *data are* stored on disk or tape on the main computer.

□ Use a singular verb when two or more subjects are joined by the word *and,* and they refer to the same person.

> *The president and chief executive officer gives* an annual goals statement to the Board of Directors. (Both nouns refer to the same person, who gives the annual statement.)

□ Use a plural verb if two or more subjects are joined by *and* or by *both...and,* and they refer to different people or things.

> *The president and the chief executive officer give* an annual goals statement to the Board of Directors. (The *president* and the *chief executive officer* refer to two different people.)

□ Make pronouns and their antecedents agree grammatically.

> Wrong: Each *user* should enter their identification number. (In this sentence, the singular antecedent, *user,* does not agree with the plural pronoun, *their.*)

FɪɢᴜʀE **8.12** ■ *(Continued)*

The tag "<h1>" indicates that a level one heading begins here, and the tag "[h1]" indicates that the level one heading should end. The writer or editor can decide later what a level one heading should look like. For one type of document, level one headings may be 12-point bold, but for another document, level one headings may be 24-point bold. These specific definitions are created in another software program and can be changed each time a document is printed.

Two examples of markup languages are standardized general markup language (SGML) and hypertext markup language (HTML). The first, SGML, is used for printed documents. The second, HTML, is used for hypertext pages—electronic documents such as online help screens or Web pages.

Designing electronic documents

Most of the techniques discussed so far in this chapter are appropriate for both paper and electronic documents. However, electronic documents require certain special considerations.

Web pages

To learn about Web page design in full, you will need to take classes or read books about this topic (see Lynch & Horton, 1999). Many organizations have specific employees, with job titles such as Webmaster or Web Designer, who are responsible for designing Web pages. In general, Web design requires you to consider some of the same items discussed so far in this chapter.

Typefaces and page layout. As one team of Web design experts notes, "[a]lthough the basic rules of typography are much the same for both Web pages and conventional print documents, type on-screen and type printed on paper are different in crucial ways" (Lynch & Horton, 1999, p. 79). For example, the computer screen displays typefaces at a much lower resolution than a printed document, making them harder to read. Also, long lines of text on a computer screen become blurred at the edges. Finally, even though you may select a special typeface for your Web site, you can't guarantee how this will appear on each individual computer screen, because different browsers and different computers project images differently.

Therefore, keep your choices of type simple and readable:

- Don't mix and match too many typefaces.
- Use sans serif type for body text.
- Don't use small type: anything under 12 point is hard to read.

Headings. On a Web page, links serve the purpose of headings. Each link takes readers to a deeper level of information. As with printed text headings, links should be consistent: Use the same typeface and font for the same level heading. Links also require user testing to be sure they actually work.

In the end, you should work with a professional Web designer to be sure your page will look and function the way you want it to.

Online help

Like Web page design, designing online help screens is a specialty. Many organizations, especially those that produce software, hire technical communicators who know how to produce online help screens. As with all page design, paper or electronic, producing online help screens requires consistency.

Typefaces. As with Web pages, online help screens are usually shaped differently from printed pages, so lines cannot be set at the same length as in a book. Sans serif type is usually more readable on the screen, and anything under 12 point will be hard to read.

Headings. The headings in an online help screen should be consistent. If help screens link to other screens, these links need to be checked to make sure they are functional.

CD-ROMs

If you are designing information for a CD-ROM, the same concepts apply. Typefaces need to be clear and legible for the screen. Different fonts should be used consistently. Headings should also be consistent, and any links should be tested.

 REVIEW CHECKLIST

Typography

Typefaces	Type Families	Uses
Serif	Times, Palatino	Body text, formal correspondence
Sans serif	Helvetica, Frutiger	Headings, technical material
Display	Optima, other decorative fonts	Headings, body copy for special effects

(continued)

- Try using fonts from only one typeface.
- If you mix different typefaces, be consistent.
- Use *italic,* **bold,** and ALL CAPS sparingly.
- Use customary formats for specific purposes.

Page Layout

- Use a consistent visual hierarchy.
- Use white space and create areas of emphasis.

Headings

- Address reader questions.
- Create headings as points of access to the text.
- Make headings visually consistent and grammatically parallel.

Other Items

- Create an effective table of contents and index.
- Use style sheets.
- Use page layout software for complicated documents.
- Pay special attention to the unique requirements of electronic documents.

Exercises

1. In this chapter, you learned that readers scan a document first in order to make sense of it and to understand the document's organizational structure and visual hierarchy. To test this idea, try the following experiment. Take a page from a document that contains enough visual cues, such as headings, subheadings, tables, and so on. Make a transparency of this document. In class, put this transparency on the overhead projector, and turn the focus knob so the document is out of focus. Make sure no one can see the actual text but that they can see the structure of the page. Ask your classmates to point out the main headings, subheadings, and areas that contain graphics. In class, discuss how people knew this information without being able to read the actual text.

2. **Focus on ▸ Writing.**   With a partner, locate two printed documents: one that demonstrates good use of typefaces, and one that demonstrates confusing or inconsistent use. Imagine that you and your partner are a team of technical communication consultants. Write a memo to the manager of the organization involved with the effective document and explain its positive

features. Write a memo to the manager of the organization involved with the confusing document and make suggestions for improvement.

3. Find a document that is intended to answer a question: a patient brochure, a reference guide for new students on campus, or a personnel document from your company. Redesign this document so it uses headings in the form of reader questions. Also, make any changes necessary to allow for the effective use of white space.

The Collaboration Window

Form teams of three or four people. Using the document you worked with in Exercise 3, appoint one person on your team to play the role of editor. The editor is responsible for making sure that formatting, typeface choices, headings, and other page design elements are used consistently.

Before you redesign the document, develop a style sheet for your team to follow. Then, ask each team member to work on one section of the document. Bring all the sections back together (either as paper printouts or on disk), and ask your editor to review your materials. The editor should check your work against the group style sheet. If some of the work is inconsistent with the style sheet, talk about ways you could improve the wording or layout of the style sheet so it is easier for writers to follow.

The Global Window

Locate a document that presents the same information in several languages (assembly instructions for various products are often written in two or three languages, for example). Evaluate the design decisions made in these documents. For example, are the different languages presented side by side or in different sections? Write a memo to classmates and your instructor evaluating the document and making recommendations for how it might be improved.

Also, talk to someone who is involved in translating documents (large cities often have translation companies, and your university or college will probably have an international student office). Ask these professionals if type or page design has any effect on how a document might be translated. For example, if a series of headings were set in boldface in the English document, how would these be set in a German, Japanese, or Italian document?

Click on This

Do a search on "fonts," "typefaces," or "page design" using your favorite search engine. Share your findings with the class, describing the kinds of sites you found and what you learned. Here are some additional sites that you may find useful:

- www.fontsite.com   *Fontsite:* A magazine for type and graphics professionals. Offers downloadable fonts, style sheets, and more.

- www.gammag.com   *Graphic Arts Monthly:* A magazine for graphics arts professionals.

- www.adobe.com   Adobe Systems: Adobe is a software company that produces popular page layout, font, photography, and other software tools.

CHAPTER **9**

Graphics and Visual Information

The power of the picture

When to use visuals

Different visuals for different audiences

Text into tables

Numbers into images

Illustrations

Diagrams

Symbols and icons

Wordless instruction

Photographs

Maps

Visualization and medical imaging

Software and web-based images

Using color

Avoiding visual noise

Visuals and ethics

Cultural considerations

Review checklist

Exercises

The collaboration window

The global window

Click on this

THE POWER OF THE PICTURE

According to expert William Horton, "[w]e all think visually" (1991, p. 16). Visual communication is a very basic form of human communication, predating written language. Before there were alphabets or symbols for numbers, humans communicated visually. More than 15,000 years ago, humans created cave paintings of animals, hunting expeditions, and other activities. Today, we are surrounded by powerful images: charts that represent the stock market; television advertisements; photos and illustrations. Visual communication remains vital.

When people look at a visual pattern, such as a graph, they see it as one large pattern—a whole unit that conveys information quickly and efficiently. For instance, the line graph in Figure 9.1 has no verbal information. The axes are not labeled, nor is the topic identified. But one quick glance, without the help of any words, tells you that the trend is rising. The graph conveys information in a way plain text never could. It would certainly be hard for audiences to visualize this trend by just reading a long list of numbers, such as:

> The stock began at 15-7/8, then rose to 16. It rose again to 17, 18, 18-1/2, and 19, then leveled off at 19 for several days. . . .

Visuals are especially important in technical communication because they enhance accessibility, usability, and relevance.

- **Accessibility.** Because humans understand visuals intuitively, visual information makes your content accessible to a wide audience. Also, if your manual or report is written in English, charts or graphs can often be easily understood by non-English speakers. For example, a report on the

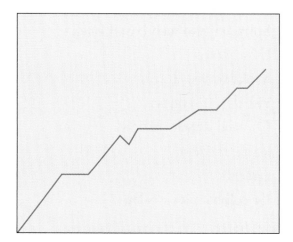

FIGURE **9.1** ■ **Line graph with no labels.**

European economy written in English might be difficult for non-English speakers to read, but a graph of European financial trends would be more broadly accessible.

- **Usability.**   Information is usable when audiences can find what they need to perform the tasks at hand. Visuals can simplify this process, because they focus and organize information, making it easier to remember and interpret. A simple table, for instance, can summarize a long and difficult passage of text. A pie chart can show the relationship between parts and a whole.

- **Relevance.**   Information is relevant if people can relate the content of the information to the task they need to perform. Sometimes a series of numbers or a long passage of text strikes readers as irrelevant. But a well-designed visual, such as a pie chart or diagram, can help readers see the connection between this information and their task or project. If your project generates 35% of company revenues, for example, viewing that slice of a pie chart may have more impact on the audience than simply reading the percentage as text.

When to use visuals

In general, you should use visuals whenever they make your point more clearly than text or when they enhance your text. Use visuals to clarify and enhance your discussion, not just to decorate your document. Use visuals to direct the audience's focus or help them remember something. There may be organizational reasons for using visuals; for example, some companies may always expect a chart or graph as part of their annual report. Certain industries, such as the financial sector, often use graphs and charts (such as the graph of the daily Dow Jones Industrial Average).

Different visuals for different audiences

Like all effective technical communication, visuals must fit your audience and purpose. For example, Figure 9.2 shows a special type of chart called a temperature surface plot. This chart takes specific pieces of data and "plots" them across a given geographic area. Such a chart makes complete sense to a trained meteorologist but would baffle a general audience. The concept ("temperature surface plot") is not something this audience would be familiar with. The axes are labeled with strange numbers that do not immediately convey any meaning. The plot itself is not a familiar shape.

Compare this chart with the line graph in Figure 9.3, taken from the National Climate Data Center Web site, a site designed for a more general audience. This graph, which shows the percentages of both wet and dry areas in the U.S. over a four-year period, is much more accessible and familiar to

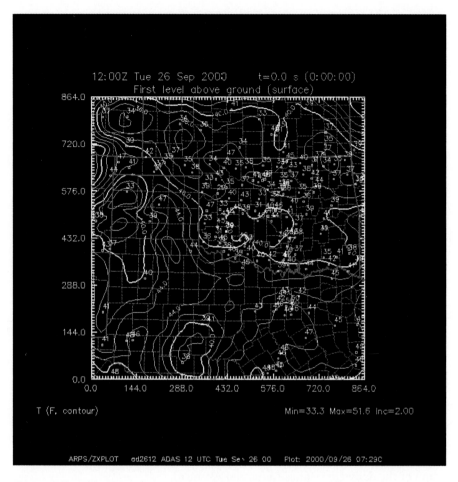

Figure 9.2 ■ **Surface temperature plot for a scientific audience.**
Source: The Center for Analysis and Prediction of Storms at the University of Oklahoma.

general audiences than the surface plot in Figure 9.2. Additionally, the axes of the line graph are labeled with dates and numbers that make sense to a non-specialized reader.

In short, a visual's content must be familiar to the audience, and the type of visual must also be understandable.

Text into tables

A table is a powerful way to illustrate dense textual information, such as specifications, comparisons, or conditions. Figure 9.4 shows a page from an instruction manual for a weather radio. In its purely textual form, this informa-

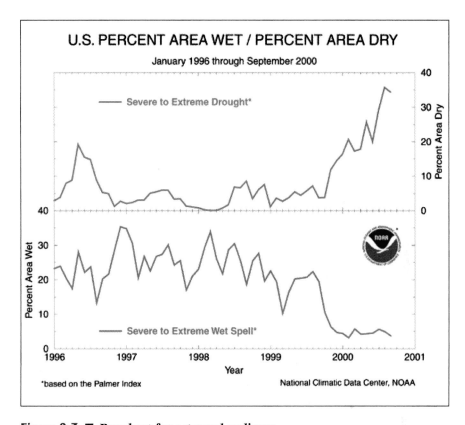

Figure 9.3 ■ **Bar chart for a general audience.**
Source: National Oceanic and Atmospheric Administration (NOAA), National Climate Data Center. www.ncdc.noaa.gov.

tion is hard to follow, because readers have to jump back and forth between the sentences to compare the different conditions. But in a table format, parallel information can be listed in the same column and row, as shown in Figure 9.5.

The table shown in Figure 9.6 explains the various types of kayak paddles and provides an overview of the specifications for each. This tabular information is much more accessible than any text-based equivalent. Readers can select the general type of paddle they are interested in: Premium Touring, for example. Readers can then determine which specific paddle (San Juan versus Little Dipper) is the length, weight, and type of material they need for their particular type of kayaking.

Numbers into Images

Visuals are especially effective in translating numeric data into shapes, shades, and patterns. Graphs and charts help you achieve this purpose.

Monitoring NOAA Signals

There are three different monitoring options: Mute, Standby and Speaker

 Note: *For all three of the following functions, the* **OFF-ON** *switch must be in the "On" position*

1. By switching the **MUTE/STANDBY/SPEAKER** switch to "Standby," your Oregon Scientific All Hazards Monitor will automatically sound the Emergency Alert Signal when it is broadcast, the speaker will automatically be turned on, and the red LED indicator will light.

2. By sliding the **MUTE/STANDBY/SPEAKER** switch to "Mute" the "ALERT" icon and red LED indicator will flash; however, the speaker will remain silent.
   To listen to the Emergency Broadcast Message, you must manually slide the **MUTE/STANDBY/SPEAKER** switch to "Speaker."

3. By sliding the **MUTE/STANDBY/SPEAKER** switch to "Speaker, " all NOAA current broadcasts will be audible.

Figure 9.4 ■ **Information from a weather radio manual in text format.**
Source: Oregon Scientific Inc., Portland, OR.

How to monitor the NOAA Weather Signals

Your weather radio has three modes: Mute, Standby, and Speaker. In order for any of these modes to function, you must first turn the power switch to "ON."

Mode	Indicator	Action	Condition
Mute	Red will flash	Speaker will remain silent	When emergency alert signal is sent by NOAA
Standby	Red will light up	Speaker will automatically turn on	When emergency alert signal is sent by NOAA
Speaker	Green will light up	All current NOAA broadcasts will be audible	Any time you wish to listen to weather broadcast

Figure 9.5 ■ **Information from the weather radio manual (Fig. 9.4) reformatted in table format.**

Specifications

Premium Touring

San Juan	full size blade
Paddle Length	220-260 by 10 cm
Blade Length	56 cm
Blade Width	17 cm
Weight for Size	230 cm
Standard	921 gr 32.5 oz
All Carbon	794 gr 28 oz
Ultra-Light	652 gr 23 oz

Camano	mid size blade
Paddle Length	220-260 by 10 cm
Blade Length	52 cm
Blade Width	16 cm
Weight for Size	230 cm
Standard	865 gr 30.5 oz
All Carbon	765 gr 27 oz
Ultra-Light	624 gr 22 oz

Little Dipper	small size blade
Paddle Length	220-260 by 10 cm
Blade Length	48 cm
Blade Width	15 cm
Weight for Size	230 cm
Standard	850 gr 30 oz
All Carbon	751 gr 26.5 oz
Ultra-Light	609 gr 21.5 oz

Kauai	mid size blade
Paddle Length	210-230 by 5 cm
Blade Length	48 cm
Blade Width	18.5 cm
Weight for Size	220 cm
Standard	865 gr 30.5 oz
All Carbon	751 gr 26.5 oz
Ultra-Light	609 gr 21.5 oz

Molokia	full size blade
Paddle Length	210-230 by 5 cm
Blade Length	49 cm
Blade Width	20.5 cm
Weight for Size	220 cm
Standard	907 gr 32 oz
All Carbon	794 gr 28 oz
Ultra-Light	638 gr 22.5 oz

Premium Whitewater

Rogue	full size blade
Paddle Length	194-203 by 3 cm
Blade Length	49 cm
Blade Width	19.5 cm
Weight for Size	200 cm
Standard	1049 gr 37 oz
Carbon Blades	1006 gr 35.5 oz
All Carbon	950 gr 33.5 oz

Quest	mid size blade
Paddle Length	194-203 by 3 cm
Blade Length	46 cm
Blade Width	18.5 cm
Weight for Size	200 cm
Standard	964 gr 34 oz
Carbon Blades	936 gr 33 oz
All Carbon	879 gr 31 oz

Freestyle	full size blade
Paddle Length	194-203 by 3 cm
Blade Length	48 cm
Blade Width	19.5 cm
Weight for Size	200 cm
Standard	992 gr 35 oz
Carbon Blades	964 gr 34 oz
All Carbon	907 gr 32 oz

Side Kick	full size blade
Paddle Length	194-203 by 3 cm
Blade Length	48 cm
Blade Width	19.5 cm
Weight for Size	200 cm
Standard	992 gr 35 oz
Carbon Blades	964 gr 34 oz
All Carbon	907 gr 32 oz

Nantahala	full size blade
Paddle Length	54-56 by 2 in
Blade Length	55 cm
Blade Width	21.5 cm
Weight for Size	58 in
Standard	737 gr 26 oz

Mid-Line

Mid-Tour	full size blade
Paddle Length	220-240 by 10 cm
Blade Length	52 cm
Blade Width	16 cm
Weight for Size	230 cm
Standard	1106 gr 39 oz

Mid-Tour S	mid size blade
Paddle Length	220-240 by 10 cm
Blade Length	48 cm
Blade Width	15 cm
Weight for Size	230 cm
Standard	1077 gr 38 oz

Mid-Sport	mid size blade
Paddle Length	210-220 by 5 cm
Blade Length	47 cm
Blade Width	18.5 cm
Weight for Size	220 cm
Standard	1077 gr 38 oz

Mid-WW	mid size blade
Paddle Length	194-203 by 3 cm
Blade Length	45 cm
Blade Width	18.5 cm
Weight for Size	200 cm
Standard	1021 gr 36 oz

Mid-WW 3 Pc.	mid size blade
Paddle Length	194-203 by 3 cm
Blade Length	45 cm
Blade Width	18.5 cm
Weight for Size	200 cm
Standard	1134 gr 40 oz

Point Paddles

Point Kids	small size blade
Paddle Length	180-230 by 10 cm
Blade Length	49 cm
Blade Width	12.5 cm

Point Canoe	full size blade
Paddle Length	54-62 by 2 in
Blade Length	54.5 cm
Blade Width	21.5 cm

Our Warranty: WERNER PADDLES are warranted to be free from defects in material and workmanship for a period of one year from the original date of purchase. Within that warranty period all paddles found to have defects will be repaired or replaced at no charge.

Figure 9.6 ■ **Kayak paddle information.**
Source: Werner Paddles, Inc.

Graphs

Graphs display, at a glance, the approximate values, the point being made about those values, and the relationship being emphasized.

Simple line graph. A simple line graph, as in Figure 9.7, uses one line to plot time intervals on the horizontal scale and values on the vertical scale.

Multiline graph. The multiline graph in Figure 9.8 uses three lines to illustrate separate trends for two major types of information technology (IT) workers, as well as the overall trend for all workers. Readers can see that while the overall trend is upwards, wages for IT workers are consistently higher than the average wage for all workers.

Band graph. The next graph, Figure 9.9, is also a type of line graph called a band or area graph. By shading in the areas beneath the main plot lines, you can highlight specific features. Despite their visual appeal, multiple-band graphs are easy to misinterpret: In a multiline graph, each line depicts its own distance from the zero baseline. But in a multiple-band graph, the very top line depicts the total, with each band below it being a part of that total. Always clarify these relationships for users.

Bar graph. Bar graphs show discrete comparisons, such as year by year or month by month. Each bar represents a specific quantity. You can use bar graphs to focus on changes in one value or to compare values over time.

Simple bar graph. A simple bar graph displays one trend or theme. The simple bar graph in Figure 9.10 is derived from United States Census data for the year

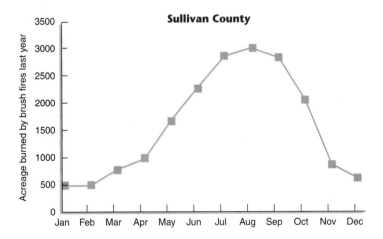

Fiɢuʀᴇ **9.7** ■ Simple line graph.

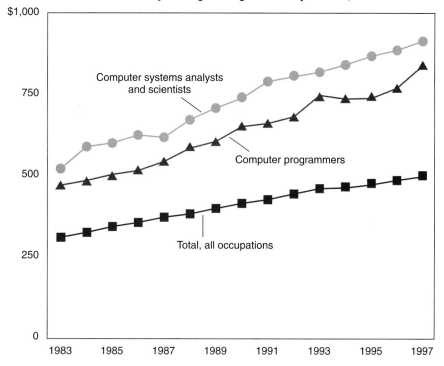

Median weekly earnings of wage and salary workers, 1983–97

FiƧURE **9.8** ■ Multiline graph.
Source: Chart prepared by the Bureau of Labor Statistics.

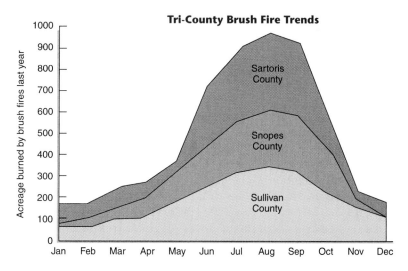

FiƧURE **9.9** ■ Multiple-band graph.

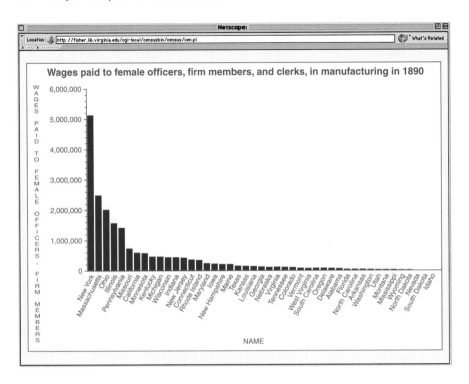

Figure 9.10 ■ **Simple bar graph.** Created on the Web using U.S. Census data from 1890.
Source: United States Historical Census Data Browser. Author/Host: University of Virginia Library Geospatial and Statistical Data Center. fisher.lib.virginia.edu

1890 and illustrates the wages paid that year to female officers, firm members, and clerks in the manufacturing field. This graph was easy to create by going to the Web site of the Interuniversity Consortium for Political and Social Research (ICPSR) at fisher.lib.virginia.edu/census. This site allows you to search through census data beginning in 1790 and generate statistical data as well as charts. If you were working on a report about the history of trends in pay scales of males versus females, you might want to create such a chart to see how each state differed over time. As with a line graph, you can clearly see a trend, but in this case, the trend is from the state that paid the highest (New York) to the one that paid the lowest (Idaho).

Bar graphs call attention to the high and low points by focusing the eye on the highest or lowest bar. Note that while this bar graph may be readable on a Web page, it is hard to read the specific labels when the graph is reduced to fit on the page of this book. You should always consider your final product when creating a bar graph or any other visual, because what is easy to understand on the computer screen may not be as clear when you turn it into a printed page, transparency, or handout, and vice versa.

Multiple-bar graph. A bar graph can display two or three relationships simultaneously. Figure 9.11 contrasts three sets of information, allowing readers to see three trends. When you create a multiple-bar graph, be sure to use a different color or pattern for each bar, and include a key so your audience knows which color or pattern corresponds with which bar. The more relationships you include on a graph, the more complex the graph becomes, so try not to include more than three on any one graph.

Deviation bar graphs. Most graphs begin at a zero axis point, displaying only positive values. A deviation bar graph, however, displays both positive and negative values, as in Figure 9.12. Note how the horizontal axis extends to the negative side of the zero baseline, following the same incremental division as the positive side of the graph.

Charts

The terms *graph* and *chart* are often used interchangeably. But a chart displays relationships that are not plotted on a coordinate system (*x* and *y* axes). Commonly used charts include pie charts, Gantt charts, tree charts, and pictograms.

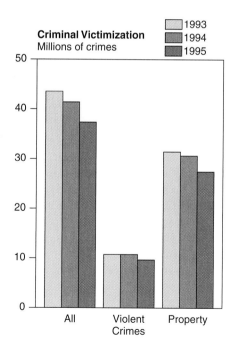

FiɢuRE **9.11 ■ Multiple-bar graph.**
Source: U.S. Bureau of the Census.

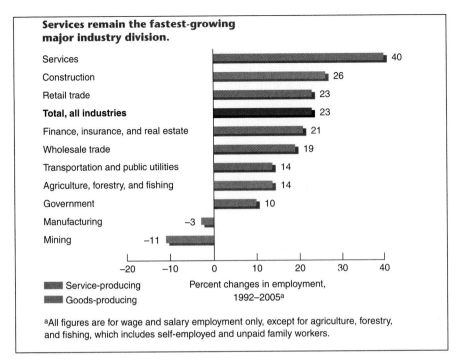

Figure 9.12 ■ **Deviation bar graph.**
Source: Bureau of Labor Statistics.

Pie chart. Pie charts are one of the most common forms of charts and are easy for almost anyone to understand. Pie charts display the relationship of parts or percentages to the whole. In a pie chart, readers can compare the parts to each other as well as to the whole (to show how much was spent on what, how much income comes from which sources, and so on). Figure 9.13 shows a pie chart created with a personal finance program. This chart makes it very clear that for this household, groceries are the largest annual expense. Figure 9.14 is an exploded pie chart created in a spreadsheet program. Exploded pie charts help call out, or highlight, the various pieces of the pie. Once you have created a chart like this, you can easily save it as an image and include it in a word-processing file (for a report) or as part of a Web page.

With pie charts, make sure the parts add up to 100 %. Use different colors or shades to distinguish between parts and the whole, or differentiate by exploding out each pie "slice." Include a key to help readers differentiate between the parts or label each slice directly. If most of the slices are very small or quite similar in size, consider using a different format, perhaps a bar graph.

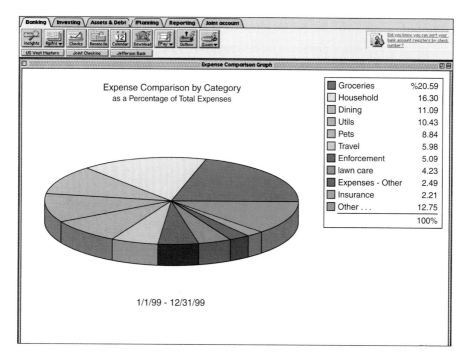

FIGURE 9.13 ■ **Pie chart.** Easy to create with a personal finance program, such as *Quicken*™.
Source: Courtesy of Intuit.

Gantt chart. Gantt charts (named for engineer H. L. Gantt, 1861–1919) depict how the parts of an idea or concept relate to each other. A series of bars or lines (time lines) indicates beginning and completion dates for each phase or task in a project. Figure 9.15 is a Gantt chart illustrating the schedule for a manufacturing project. Many professionals use project management software to produce Gantt and similar charts (more on this later, in Software and Computer Graphics).

Tree charts. Many types of charts can be generally categorized as "tree" charts. These include the following:

- flowcharts, which use a tree structure to trace a procedure from beginning to end
- software charts, which use a tree structure to outline the logical steps in a computer program
- organization charts, which show the hierarchy and relationships between different departments and other units in an organization

Figure 9.16 shows an organizational tree chart.

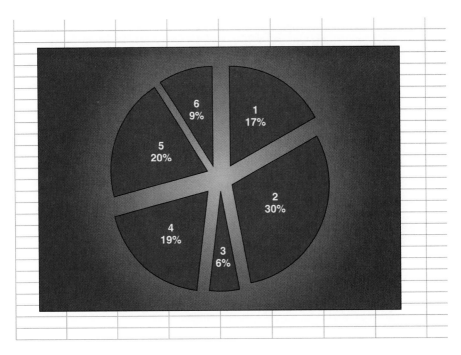

Figure 9.14 ■ **Exploded pie chart.** Custom color background created in *Microsoft Excel*™.
Source: Reprinted by permission from Microsoft Corporation.

	Activity	Start Date	Finish Date	1996										Actual Duration
				August				September				October		
				5	12	19	26	2	9	16	23	30	7	14	
1	Design														19.10
2	Brainstorming	8/11	8/18												7.14
3	Research	8/18	8/23												4.98
4	Marketing	8/23	8/30												6.98
5	Testing														31.06
6	Prototype	8/13	8/25												11.88
7	Drawings	8/25	9/2												8.28
8	Build	8/25	9/14												19.60
9	Monthly Reviews														
10	Production														50.96
11	Factory Prep	9/17	10/7												20.30
12	Materials Delivery	10/7	10/21												13.30
13	Production	10/21	11/2												12.18
14	Begin Shipping	11/7													

Figure 9.15 ■ **Gantt chart.** This chart shows the schedule for a manufacturing project.
Source: Courtesy of AEC Software, © 1996.

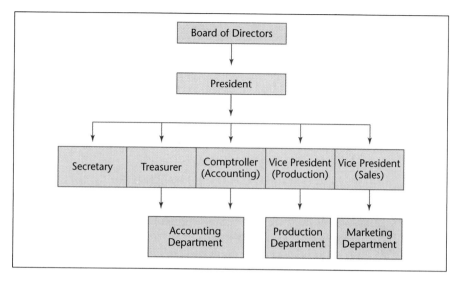

FiGure 9.16 ■ Organizational chart.

Pictogram. Pictograms are something of a cross between a bar graph and a chart. Like line graphs, pictograms display numerical data, often by plotting it across an x and y axis. But like a chart, pictograms use icons, symbols, and other graphic devices rather than simple lines or bars. Figure 9.17 is a pictogram that uses stick figure icons to illustrate population changes during a given period. Pictograms are visually appealing and can be especially useful for nontechnical audiences. Graphics software makes it easy to create pictograms.

Illustrations

An illustration is sometimes the best and only way to convey information. Illustrations can be drawings, diagrams, symbols, icons, photographs, maps, or any other visual that relies on pictures rather than on data or words. For example, the drawing of the brain in Figure 9.18 accomplishes what plain text cannot: It offers an overview of the brain's shape, the relative sizes of its segments, and its structure. Illustrations are invaluable when you need to convey spatial relationships or help your audience see what something actually looks like. Drawings can be more effective than photographs, because in a drawing, you can simplify the view, remove any unnecessary features, and focus on what is important.

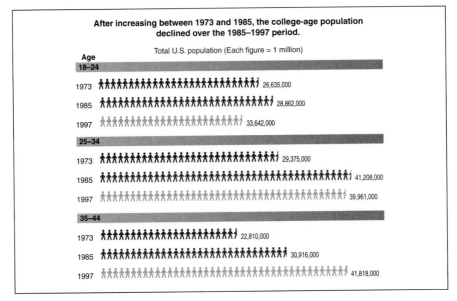

FIGURE 9.17 ■ Pictogram.
Source: U.S. Bureau of the Census.

DIAGRAMS

Diagrams are a useful way to illustrate a device or part of a device. Figure 9.19 is a diagram of a cloud-simulation chamber that provides readers with an understanding of the device as a whole and of the parts that make up the device.

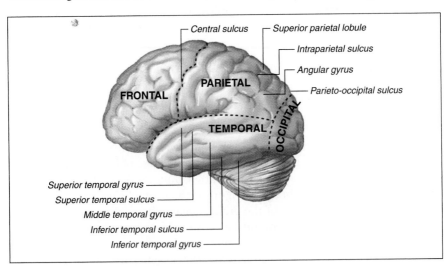

FIGURE 9.18 ■ An illustration of the brain.
Source: Davidoff, J. (1991). *Cognition through color.* Cambridge: MIT Press.

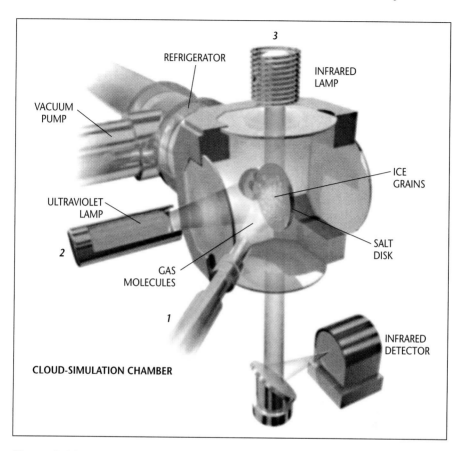

Figure 9.19 ■ **Cloud simulation chamber.** Diagrams are useful for illustrating devices.

Source: Bernstein, M. P., et al. (1999, July). Life's far-flung raw materials. *Scientific American,* 42–49.

Exploded diagrams

An exploded diagram can help you explain how a component fits or how a user should assemble a product. The exploded diagram in Figure 9.20 illustrates how to assemble a field mill, which measures fluctuations in the earth's electrical field.

Cutaway diagram

Cutaway diagrams are extremely useful for showing your audience what is inside of a device or to help explain how a device works. The cutaway diagram in Figure 9.21 illustrates how fireworks operate.

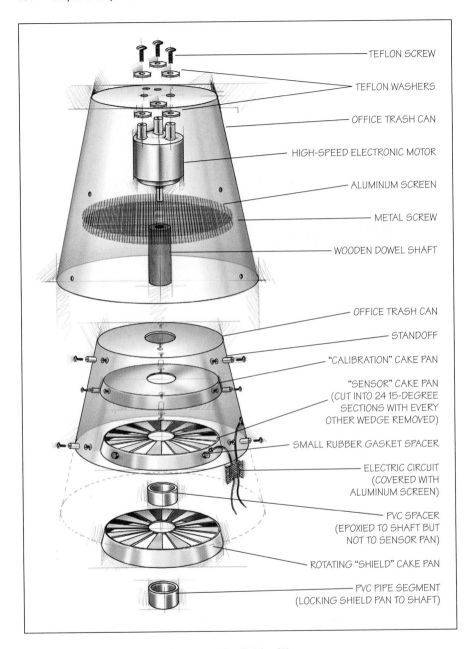

Figure 9.20 ■ **Exploded diagram of a field mill.**
Source: Carlson, S. (1999, July). Detecting earth's electricity. *Scientific American,* 94–95.

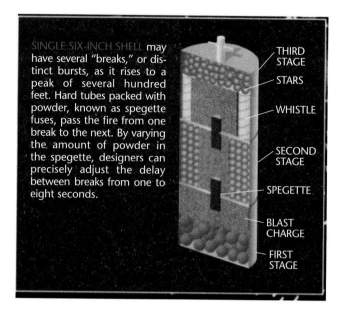

SINGLE SIX-INCH SHELL may have several "breaks," or distinct bursts, as it rises to a peak of several hundred feet. Hard tubes packed with powder, known as spegette fuses, pass the fire from one break to the next. By varying the amount of powder in the spegette, designers can precisely adjust the delay between breaks from one to eight seconds.

THIRD STAGE
STARS
WHISTLE
SECOND STAGE
SPEGETTE
BLAST CHARGE
FIRST STAGE

Figure 9.21 ■ Cutaway diagram.
Source: Zambelli, G. R., Sr. (1999, July). Aerial fireworks. *Scientific American*, 108.

Symbols and icons

Symbols and icons are useful ways to make information available to a wide range of audiences. Because such visuals do not rely on text, they are often more easily understood by international audiences, children, or people who may have difficulty reading. Symbols and icons are used in airports, shopping malls, restaurants, and other public places. They are also used in documentation, manuals, or training material, especially when the audience is international. Some of these symbols are developed and approved by the International Standards Organization (ISO). The ISO makes sure the symbols have universal appeal and are standard, whether used in a printed document or an elevator wall.

The words *symbol* and *icon* are often used interchangeably. Technically, icons tend to resemble the thing they represent: an icon of a file folder on your computer, for example, looks like a real file folder. Symbols can be more abstract; symbols still get the meaning across but may not resemble, precisely, what they represent.

Figure 9.22 shows some international symbols used in airports and other public places. You should have no difficulty guessing what each means.

Note that the first three symbols represent nouns: things or objects. The last symbol (the arrow), however, represents a verb. It indicates that the person

Fɪɢᴜʀᴇ 9.22 ■ Internationally recognized symbols.
Source: 4YEO.com, www.4yeo.com/ICONS/signs/index.htm

looking at it should do something—in this case, turn right. Nouns are easier to represent than verbs, because drawing a thing is easier than drawing an action.

If you are creating a brochure, Web site, manual, or other information product for an international audience, consider using some ISO symbols. Check the ISO Web site at www.iso.ch and related sites, which you can locate using a Web search engine.

Wᴏʀᴅʟᴇss ɪɴsᴛʀᴜᴄᴛɪᴏɴ

Many organizations with international audiences rely on wordless instruction: symbols, drawings, and diagrams that convey information completely without words. Flight information cards in the seat pockets of commercial airplanes are an example of wordless instruction, because they use diagrams, photographs, and drawings to explain how passengers can exit an aircraft in the event of an emergency. Figure 9.23 shows how the manufacturer of an automatic coffeemaker uses wordless instruction for its international audience.

Pʜᴏᴛᴏɢʀᴀᴘʜs

Photographs are especially useful for showing what something looks like. Unlike a diagram, which often highlights certain parts of an item, photographs show everything. So, while a photograph can be extremely useful, it can also provide too much detail or fail to emphasize the parts you want your audience to focus on (see Figure 9.24). To obtain the most effective photograph, use a professional photographer who knows all about angles, lighting, lenses, and special film.

Mᴀᴘs

Besides being visually engaging, maps are especially useful for showing comparisons and helping users visualize the position, location, and the relationship between various data. The map in Figure 9.25 synthesizes information about various states, regions, and agricultural product values in the U.S.

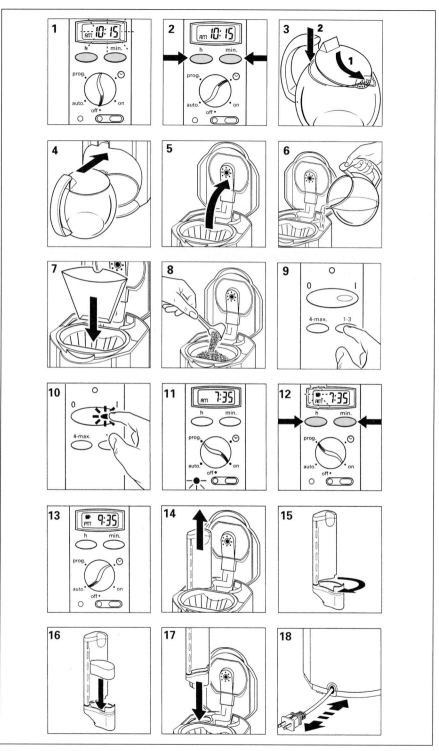

Figure 9.23 ■ **Wordless instructions.** This coffeemaker owner's manual uses wordless instruction.

Source: Courtesy of Krups, Inc.

Figure 9.24 ■ **Photographs.** Photographs provide a good view of the entire product, such as this electronic car.
Source: Capital Features/The Image Works.

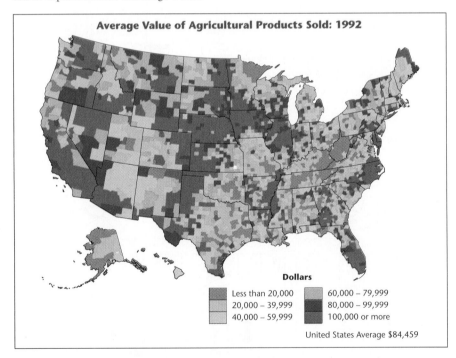

Figure 9.25 ■ **A Map.** Agricultural product values in the United States.
Source: U.S. Department of Commerce.

Visualization and medical imaging

Certain techniques from science and medicine can provide powerful and useful visuals for technical communication. Visualization is a process whereby scientists load complex mathematical data into a high-speed computer, and the computer generates an image. The image becomes a tool that can be used to understand the data. For example, images of the surface of Mars are sent back as numbers. Complex software then takes this numerical information and creates an image. The structure of DNA, which is also represented mathematically, can be converted into a visual image that scientists use to envision how DNA works.

Similarly, medical images such as CAT scans, ultrasounds, and photographs of the inside of the body via a laparoscopic camera give us a visual picture of organs, muscles, and tissue not visible from outside the body. These images, which are often available on the Web, might be appropriate in a medical textbook or a manual used by physicians to learn about a new imaging product.

Software and Web-based images

You can create most of the visuals discussed in this chapter with the wide assortment of computer graphics software, spreadsheets, presentation software, and related products availiable today. Also, you can find clip art, icons and symbols, medical images, and a vast assortment of other visuals on the Web. (See Chapter 7 for information on copyright and Web-based graphics.)

Consider taking a class in computer graphics. With the continued growth of the Web, the Internet, cable television, personal computing, and other visually based technologies, anyone's career can be enhanced by learning about visuals and computer technology.

Knowing about the following categories of computer software will definitely enhance your abilities as a technical communicator—whether this is your full-time job or just part of your job.

- *Graphics software*, such as *Adobe Illustrator* or *CorelDraw*, allows you to draw and illustrate.
- *Photography software*, such as *Adobe Photoshop*, allows you to work with photographs, scanned images, or other files (such as graphics files taken from the Web). You can manipulate these images to make them fit your document or to make them more appropriate for your audience and purpose. But don't change the items so much that you change the meaning.
- *Presentation software*, such as *Microsoft PowerPoint*, lets you create slides, computer presentations, and overhead transparency sheets. Other types of presentation software, such as *Macromedia Director*, are designed to create complex multimedia presentations that include sound and video.

- *Project management software*, such as *Microsoft Project,* make it easy to create Gantt charts or organizational tree charts.
- *Spreadsheet software*, such as *Microsoft Excel,* make it simple to create pie charts, bar graphs, line graphs, and so on.
- *Word-processing programs*, such as *Corel WordPerfect* or *Microsoft Word,* include simple image editors ("draw" feature) and other tools for working with visuals. More sophisticated page layout programs, such as *Adobe InDesign,* also provide ways for you to work with visuals.

Using color

Years ago, it was expensive and difficult to use color in anything but the most high-end documents. Adding color often meant adding cost, because color usually involved cleaning the printing press, using new ink, and then cleaning the press again for a different color job. Smaller color projects, such as overhead transparencies, were often colored by hand. Today, color computer screens and inexpensive color ink jet printers can make the task, especially for small jobs, far easier and cheaper. Color is an effective tool in a visual, because it helps focus reader attention and makes the document more visually interesting. Color can help you organize your visual. It can also help you orient your reader and emphasize certain areas of your visual or document. Yet too much color, or the wrong color combinations, can be worse than no color at all. Color can greatly enhance how audiences understand technical data, but there is still much we can learn about how people perceive color.

Figure 9.26 illustrates an effective use of color on the Web site for Best Buy, an electronics company. Best Buy uses yellow on its storefront signs, and the Web page continues this theme. The left side bar is yellow, and so is the store name and sign. The side bar contains links to product categories and other information. The word "online" is in orange, which complements the yellow. The page does not overuse the yellow.

When using color, remember the following items:

- **Use colors consistently.** Yellow (or any color) should mean the same thing throughout the visual or the entire document.
- **Use colors selectively.** Too many colors create "visual noise" (see below). Don't mix and match too many colors, and use colors to help organize the page. Avoid using light colors on a light background: Darker print provides more contrast with a light background.
- **Consider audience and purpose.** Certain audiences and situations may call for certain colors. For example, if your document addresses university students, you might wish to use the school colors. Also, not all audiences will see color as you do. Approximately 9% of people have color-

Figure 9.26 ■ A Web site that uses color effectively.
Source: Best Buy, Inc. www.bestbuy.com

deficient vision (Fortner & Meyer, 1997, p. 58), which prevents them from discriminating between certain shades or colors. Whenever possible, test your color choices on audience members.

- **Consider intercultural issues.**   Colors have different meanings in different countries. In China, red may mean prosperity, while in the United States, red often indicates danger (Hoft, 1995, p. 267).
- **Consider the medium.**   It is cheaper and easier to add color to a Web page or *PowerPoint* presentation than to use color on a glossy brochure. Consider how your document will be published before making color choices.

Avoiding visual noise

Too many visuals, or visuals that are crowded with too much information, create "visual noise." People have an easy time processing visuals, but not if the chart, graph, or other visual is so crowded or disorderly that it cannot be understood. One expert refers to this as "chartjunk" (Tufte, 1990, p. 34): too many bells and whistles at the expense of readable, credible visual information.

Visuals that look fine in one format may appear crowded in another. Look back at Figure 9.10. On the original Web version, the labels of state names were easy to read, because they were displayed on a browser (*Netscape* or *Explorer*) and filled an entire computer screen. Yet on the printed page, especially when reduced in size, what was once an effective visual becomes crowded and hard to read. Keep this in mind when you move from one medium to another.

Also, when creating visuals use a minimalist approach. Keep your designs simple and elegant. Don't use too much clip art, too many colors, or too many images. Test your visuals to be sure they make sense to your audience. If your organization has an in-house graphic designer, ask that person's opinion.

Visuals and ethics

Visuals can manipulate as well as inform, and with currently available computer software, it is easy to create misleading visuals. When bar charts use pictures, not just bars (see "Pictograms" earlier in this chapter), for example, the relative size of the bar and type of picture might convey a particular bias (Kostelnick & Roberts, 1998, p. 292). Or, a brighter color on one bar and lighter colors on another might prevent a reader from seeing all the data equally.

Graphs can also be confusing or misleading if the axes are not labeled. Readers assume that an axis begins at zero, but this may not always be the case if an image is not labeled. Image labels can also be confusing if they are too small for readers to see; or, worse yet, there may not be any labels at all. Photographs and other images can easily be manipulated using software. Copyright (Chapter 7) is also a consideration, because an image you obtain on the Web, for example, may be protected by copyright.

Aim for fairness and accuracy. Label the axes of charts and graphs. Indicate the source of your data. Use color appropriately and evenly. If your visual accompanies text, indicate how this visual supports the textual information.

Cultural considerations

Although groups such as ISO have tried to standardize a wide range of visual information across cultures, you will not always use ISO symbols for your visuals. Visual communication does have global appeal, but charts, graphs, tables, and other visual forms are not universal. As one expert notes, "Visual communication can make cultural assumptions that are inappropriate or offensive" (Hoft, 1995, p. 4). Not all cultures read from left to right, so a chart that is supposed to be read left to right that is read in the opposite direction could be misunderstood. Color is also a cultural consideration. Representations of images that might be used in a computer icon or diagram, for example, may be culturally offensive if they portray sensitive images, such as certain animals (the cow in Hindu culture) or gender roles (women with bare heads in certain Arabic cultures).

Review Checklist

Summary of visual forms

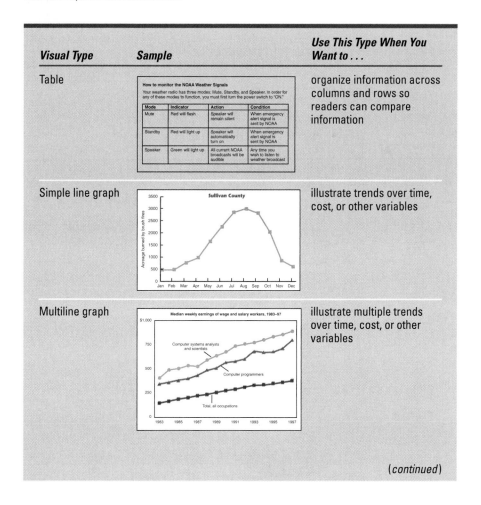

Visual Type	Sample	Use This Type When You Want to . . .
Table		organize information across columns and rows so readers can compare information
Simple line graph		illustrate trends over time, cost, or other variables
Multiline graph		illustrate multiple trends over time, cost, or other variables

(continued)

Visual Type	*Sample*	*Use This Type When You Want to . . .*
Multiple-band graph	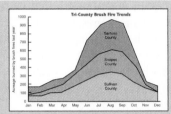	illustrate multiple trends and highlight specific features
Simple bar graph	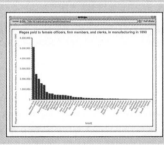	display a trend
Multiple-bar graph	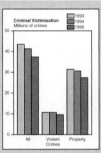	display more than one trend and their component parts
Deviation bar graph	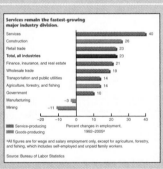	display both positive and negative values of a trend
Pie chart		display the relationship of parts or percentages to the whole

Visual Type	Sample	Use This Type When You Want to ...
Exploded pie chart	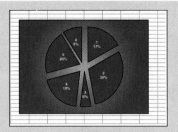	call out, or highlight, the various pieces of the pie
Gantt chart	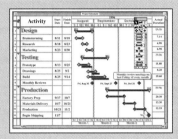	depict how the phases of a project relate to each other
Tree chart		show the flow in a program or a process, or the hierarchy in a company
Pictogram	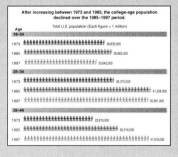	use icons, symbols, and other graphical devices (rather than simple lines or bars) in a graph or chart
Illustrating (drawing)	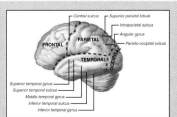	convey spatial relationships or help your audience see what something actually looks like

(*continued*)

Visual Type	Sample	Use This Type When You Want to . . .
Diagram	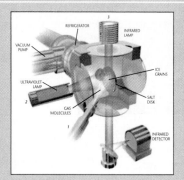	provide readers with an understanding of a device as a whole and the parts that make up the device
Exploded diagram	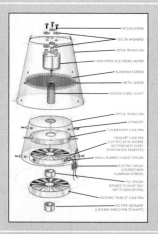	explain how a component is put together, or how a user should assemble a product
Cutaway diagram	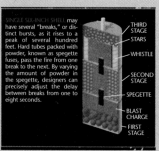	show what is inside of a device or help explain how a device works
Symbols and icons		make concepts understandable to broad audiences, including international audiences, children, or people who may have difficulty reading

Visual Type	Sample	Use This Type When You Want to . . .
Wordless instructions		convey information completely without words
Photographs		show exactly what something looks like
Maps		show comparisons and help users visualize the position, location, and the relationship between various data
Visualization and medical imaging	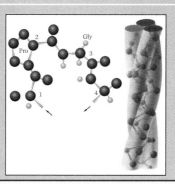	create images out of highly abstract mathematical data

EXERCISES

1. Find an article in a newspaper, journal, or magazine that does not include necessary visuals to support the communication purpose or meet the audience's needs. Analyze the article and identify where visuals would be helpful. Create a list of the visuals you would recommend to the article's author or the publication's editor.

2. From the list you created in Exercise 1, select two proposed visuals and create them using the graphics component of a word-processing program or graphics software. In a presentation to your class, explain how additional visuals would improve the usability of your article, and show the visuals you created.

3. FOCUS ON ⟫WRITING. Looking at printed materials and sites on the Web, find an example of each of the kinds of visuals discussed in this chapter (e.g., diagram, table, pie chart, simple bar chart, stacked bar chart, etc.). Critique each visual: Does it convey the information effectively? Would a different type of visual be more effective? Is visual noise distracting? What would you do to make the visual more effective, or more appropriate to audience and purpose? Write up your findings as a short memo to your instructor.

4. Obtain an instruction manual for a piece of technology with which you are familiar. Before looking at the manual, list the visuals you think might be included in the manual and consider the role each of those visuals would play in making the manual usable. Then compare your list with the visuals that are actually in the manual and analyze the differences between the manual and your list:

   - For visuals that are in the manual and on your list, compare the role in the manual and the role that you defined.

   - For visuals in the manual but not on your list, evaluate the role and effectiveness of the visuals. Do you agree with the decision to include those visuals in the manual? Why or why not?

   - For visuals not in the manual but on your list, evaluate the role that you defined, and the effectiveness of that segment of the manual as it exists without the visual. Do you agree with the decision not to include those visuals in the manual? Why or why not?

The Collaboration Window

Return to the teams you worked in for the collaborative exercise at the end of Chapter 2 and examine the audience and purpose statements you created for the Survival Guide for incoming students. (If you have not done the exercise in Chapter 2, review it now.) List the topics you are likely to include in the Guide. Then list the visuals that you believe would effectively support the purpose of each section of the Guide. As a team, present your plans to the class and explain the rationale for each proposed visual.

The Global Window

The International Standards Organization (ISO) is a group devoted to standardizing a range of material, including technical specifications and visual information. If you've ever been in an airport and seen the many international signs directing travelers to the restroom or to the smoking area (or informing them not to smoke), you have seen ISO signs. Go to the ISO Web site at www.iso.ch to learn about ISO symbols. Prepare a short report and presentation for class explaining how these symbols are developed. Show some of the symbols and explain why these work for international communication.

Click on This

Numerous Web sites allow you to download copyright-free clip art, diagrams, and other visuals that you can use in your own communications. Using a search engine, locate sources of copyright-free clip art and graphics, and also any other useful pages on visual communication. If your class has access to a Web server, create a Web page that lists these sources. Annotate your listings so that other users will know how useful a given site is. For example, some pages say they offer free clip art but offer only a few graphics for free and charge for others. Some pages offer free art, but the images are outdated or of poor quality. If your class does not have Web access, create a short handout listing the resources you found.

PART 2

Technical Communication Situations and Applications

chapter 10
Everyday Communication Situations

chapter 11
Product-Oriented Communication Situations

chapter 12
Complex Communication Situations

COMMUNICATION SITUATIONS

The next three chapters describe a variety of communication situations and include information about approaches and formats that are, in most cases, appropriate for these situations. Depending on your particular profession, company, and unique communication situation, these categories may not always be a perfect fit. Ultimately, all situations are unique, and your choices should be driven by a clear and thorough understanding of your audience, purpose, and the company or organization. For example, a situation that might call for a long report in one company might be handled by a memo or short report in another. One organization might produce all of its user documentation in printed books, while another might place all of this information on a Web site. And, in some companies, email is the preferred method of communicating, while in others, a combination of email and paper memos is the norm.

Moreover, in actual workplace applications, the categories in these chapters usually overlap. As described here, they are meant to allow you to practice discrete forms of writing and communication: writing a definition or description, writing a memo, writing a set of brief instructions. But on the job, you may need to write a memo that contains some specifications and some description, or you may need to write a short report that contains some documentation and proposes a set of actions for the company to take.

Use the next three chapters as opportunities to practice and explore. In class, discuss any experiences you may have had on the job or during an internship that differ with what you read in this book. Consider creating a class Web site that lists the most common communication situations you and your classmates encounter, and then list the types of documents you most frequently work on.

Here are the situations described in the next three chapters:

Everyday. Some situations require correspondence that takes place on a regular basis in most organizations (such as email or memos).

Product-oriented. Other situations require you to describe, explain, document, or market a technical product or service.

Complex. Still other situations require more complex approaches, which may include long reports or proposals.

Each chapter ends with a set of exercises. The Web site for this textbook (www.ablongman.com/gurak) contains links to sample documents online as well as exercises related to different document types.

No matter what the circumstance or the document, certain guidelines are always appropriate:

- Perform a thorough audience and purpose analysis.
- Create your product with usability in mind.
- Carefully select a visual format and medium.
- Consider copyright issues.

CHAPTER **10**

Everyday Communication Situations

Email

Memos

Letters

Short reports

Oral communication

Review checklist

Exercises

The collaboration window

The global window

Click on this

Email

The situation

Email has become a major form of business and technical communication, surpassing much of what used to be accomplished with paper letters and memos. Unlike paper, email can be used to quickly and efficiently address an individual, a group within an organization, or a group of interested users from outside the organization. It can reach thousands of readers in a matter of seconds, and these readers can continue forwarding the email message to others. Although paper documents can be photocopied and redistributed, few people take the time to do so. But with email, all it takes is one or two simple keystrokes, and the message has been forwarded.

Email is extremely useful in situations where people are in different time zones or have different working schedules: You can send an email at 2:00 a.m. if you are a night owl, and your early-bird colleague can read it in the morning. Email is useful if you want an electronic paper trail to track the communication—useful for helping you remember details about a project and for legal reasons, too.

Audience and purpose analysis

Unlike paper documents, with email, you don't have much control over the final audience. You may intend for your note to reach only a small group of people, but since email is so easily forwarded, your audience could turn out to be much larger. Because people tend to be more casual and off-the-cuff on email, sometimes more so than they would be in person, audience consideration becomes crucial. Suppose, for example, that after a long week of work on a particularly tough engineering design project, you send a quick email message to another manager. In your message, you complain about one of the engineers not holding up his end of things. You quickly press "Send" and head out the door. On Monday, at a status meeting, you are surprised to find that your message was forwarded to several other engineers, and they're talking about it! Clearly, your original message was not intended for them, but either by accident or on purpose, someone forwarded it along.

So, a cardinal rule about audience, purpose, and email: Always assume that your message will go far beyond its original recipient, and don't send anything on email that isn't appropriate for a wider audience or that would make you uncomfortable if used for a different purpose.

Types of email

Email messages usually follow the same basic format, with a memo style heading (Subject, Date, From, To), but often use salutations (Dear Laura) and closings (Thanks, Sam) like a letter. Email can be used for a variety of purposes

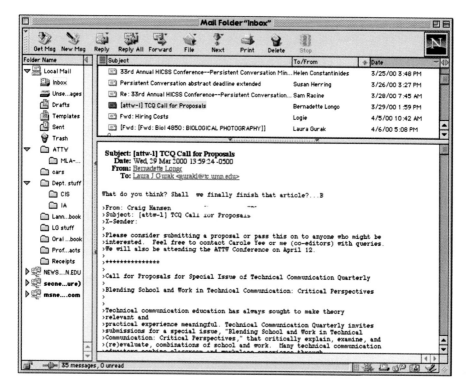

Figure 10.1 ■ An email message in *Netscape Mail*.

Source: Netscape Messenger screenshot © 2000 Netscape Communications Corporation. Used with permission.

and situations: to exchange information, share ideas, schedule meetings, and collaborate on work projects (see Figure 10.1).

Typical components of email

Most email messages have the following parts:

Header information. This information includes the standard fields from a memo (Subject, Date, From, To). Also, depending on the email program, the header may include more detailed information, such as the following:

- *host server:* the computer that generated the message
- *time of receipt:* time the computer received the message
- *return path:* an email address the computer can use if the message bounces (fails to reach its destination)
- *message ID #:* a number that the network uses to track the individual message

The header also includes a list of other recipients who received the message as a courtesy copy ("cc.")

Message body. The text of the message itself.

Attachments. Formatted documents, software, photographs, scanned images, and other files can be sent as attachments to an email message.

Usability considerations

Keep it short. Readers are impatient and don't want to scroll down through long screens full of information. Also, unbroken text is hard to read on the screen. So keep email messages brief and to the point. If you need to, put larger, related files of information on a Web page or use an attachment, where readers can read at their leisure.

Break up the information. Don't write one long block of text with no paragraph breaks. Use numbered lists, headings, and indents to break up the message.

Include links to Web sites. Include links to other relevant Web sites when necessary.

Don't send huge attachments without checking first. Not all email browsers are capable of handling attachments (formatted files, photos, images). Also, if recipients have slow Internet connections or pay by the minute for connect time, they will be annoyed by large attachments that take forever to download. Before sending an attachment, inquire as to what file types the recipient's equipment can handle and if his or her browser can accept attachments.

Pay attention to spelling and formality. Email users lean toward an informal writing style. Even though many email programs contain spell checkers, people rarely use them. Yet a correctly spelled message will be more credible than a sloppy one. Also, when composing a message for someone you don't know or someone in a position of authority, begin with a formal salutation ("Dear Professor May").

Avoid flaming. *Flaming* refers to email that is unnecessarily angry and makes extreme personal attacks. Because email users are not communicating face to face, they sometimes use email to let off steam. Always read over your message, especially in a tense or delicate situation, before sending it. Strive for messages that are informative and useful, not emotionally charged.

Don't use ALL CAPS. All capital letters are hard to read. And, in email, they have a special meaning: All caps is the equivalent of shouting. So don't shout unless you mean to.

Use smiley faces sparingly. Smiley faces, made from a colon, dash, and right-hand parenthesis :-) are used to convey humor. Use these and other emoticons infrequently, and don't use them in a formal message or in a message to someone you don't know.

MEMOS

The situation

People in almost every profession use memos to communicate business or technical information within the company, to outside vendors, to clients, or to other relevant parties. Today, the paper memo is rapidly being replaced by email. Even so, paper memos are often called for, especially if the situation is more formal or the discussion is somewhat lengthy. Use a memo when you need to

- Transmit information to a group
- Make a short evaluation or recommendation
- Distribute minutes from a meeting
- Provide follow-up on a discussion

Different organizations have different standards and practices for memos. For example, at some companies, memos are used only to convey formal information and are always written on letterhead. Other, less formal information is sent via email.

Audience and purpose analysis

To determine the length, content, tone, and approach of a memo, pay attention to the various audience members who will receive it. Some companies use standard memo distribution lists: a list for all managers, a list for all software developers, and so on. The purpose of your memo should be clear: Is it to inform your audience? To persuade them? To convince them to take action? The organization and writing style of the memo should match this purpose.

Types of memos

Memos cover every conceivable topic. See Figure 10.2 for one example. Common types are described below.

Memo of transmittal. Like a letter of transmittal, a transmittal memo is used with a package of material, such as a report, a manuscript, or a proposal. The transmittal memo introduces the material, explains what is enclosed, and of-

---

GREENTREE BIONOMICS, INC.

MEMORANDUM

TO: D. Spring, Personnal Director
FROM: M. Noll, Biology Division   *MN*
DATE: April 18, 2001
RE: Need to hire more personnel

With twenty-six active employees, Greentree has been unable to
keep up with scheduled contracts. As a result, we have a contract
backlog of roughly $500,00. This backlog is caused by understaffing
in the biology and chemistry division. To increase our production
and ease the workload, I recommend that we hire three general lab
assistants.

I have attached a short report outlining the cost benefits of these
hires. Could we meet sometime next week to discuss this in detail? I
will contact you on Monday.

cc: E. Bragin, Chemistry Division

Fιgure 10.2 ■ **Sample memo.** A memo informing employees about a workplace
hiring situation.

fers short comments or information not provided in the document itself.
Transmittal memos may be as simple as a sentence or bulleted list describing
what is enclosed.

Meeting minutes. Memos are often used to transmit minutes from a meeting. If
the minutes go beyond two pages, the memo is probably not the best format,
and you might want to opt for a short report instead.

Brief report. The memo format is often used as the basis for very brief reports.
Instead of writing a longer report with great detail, the writer will create a

short, concise version, using the basic template of a memo structure. You will learn more about short reports later in this chapter.

Typical components of a memo

Memos differ across organizations and professions, but most paper memos contain the following parts:

Name of organization. Most paper memos are printed on company letterhead, so you won't need to actually insert the name or address of the organization.

The word "memo" or "memorandum." This can be centered or set flush with the left margin.

To, from, subject, date. These four fields are set flush with the left margin and followed by a colon.

> To: Julian Barker
> From: Riley O'Donnell
> Subject: Internal price list
> Date: 15 January 2000

Sometimes, the abbreviation "Re" is used in place of "Subject." "Re" stands for "In reference to." Also, these fields may be used in a different order depending on the particular situation. In some companies, the order may be Date, To, From, Subject.

Memo body. The body copy should be set in several short paragraphs. Use lists to display prices, specifications, or other features. Use the direct approach pattern (see Figure 10.6 on page 198) whenever possible, and get to the point quickly.

Typist's initials, distribution and enclosure notations. These items are described under "Letters" and are used in a similar manner with memos.

Usability considerations

Distribute the memo to all the right people. With email this is easy because you can send to an electronic list of names. With a paper memo, however, each copy needs to be placed in a mailbox or mailed in an envelope. You may need to ask for assistance with this task; office staff often have preaddressed labels you can use.

Put the important information in an area of emphasis. Don't bury the important information in a large block of text midway through the memo. Preview the key point in the subject line, and follow the top-down approach if possible, presenting the important information clearly in the first paragraph. (See also Figure 10.6.)

Keep it short and to the point. Unless you are writing a short report in memo format (see Short Reports), keep your memo as brief as possible, to 1 page if you can.

Check spelling, grammar, and style. Run the spelling and grammar checkers, but also proofread or ask a colleague to proofread.

Make sure all appropriate parties receive a copy. No one appreciates being left out of the loop, so be sure to include everyone who should be informed.

LETTERS

The situation

Although many people use email in place of letters, traditional paper letters are still an important part of technical communication. Letters are used to address a single individual, a committee, or an organization. Letters can be short or long, depending on the context. And letters tend to be more formal than memos, email, spoken communication, or voice mail messages, because letter writers tend to be more careful than they would be with electronic communication. A letter takes time to write, print, and proofread, and in that time, a writer may decide to make a few changes or perhaps to not even send the letter at all, whereas email is quicker to compose and send.

Use a letter when you need to

- Explain something you've enclosed, such as your résumé, a report, or a manuscript
- Inquire about a product, service, or organization
- Complain about or praise a product or service
- Request technical information

Different professions have different standards for letter writing. In most engineering and science professions, letters are generally short and to the point. In certain circumstances a longer letter may be required to provide the appropriate amount of information.

Audience and purpose analysis

To determine the length, content, tone, and approach to a letter, pay attention to audience and purpose. Imagine, for example, that you are a technical writer working on a new Web site, and you order updated font software. After installing the software, you discover that it does not perform as promised. The company from which you purchased the software has a strict no refund policy on opened software. You are rather angry about the situation, so you decided to write a letter to the software manufacturer.

Who will be the audience for this letter? What is the primary purpose of your correspondence? Too often, people in this situation don't consider that their actual purpose is to obtain a refund, not to express anger. Consider the following opening:

> When I learned that your software does not perform as promised, and when I could not return it to the store, I was furious. How could you sell such a defective product?

Instead of making the recipient defensive, try to establish common ground and show that you are a reasonable person seeking a reasonable solution:

> I have always been a loyal user of your company's products, so I was disappointed that this software did not perform as stated in your marketing material. Because CompCity does not allow software returns, I ask that you refund the price. I will be glad to send the product back to you.

Along with your choice of words and tone, consider the following:

- Address your letter to the correct person(s)
- Understand how that person will feel about your request. If the person receives hundreds of letters a day, for example, he or she might feel short-tempered and might appreciate a friendly—but determined—tone
- Make sure other interested parties are copied on the letter

Types of letters

There are many types of letters including the following:

Letter of transmittal. The transmittal letter accompanies a package of material, such as a report, a manuscript, or a proposal. Usually more formal than a transmittal memo, and addressed to a recipient outside your organization, the transmittal letter introduces the material, explains what is enclosed, and offers any additional comments or information not offered in the document itself. Transmittal letters often include a sentence describing what is enclosed, such as:

> This package contains a 12-page proposal, a price list, and my business card.

Make sure you address your transmittal letter to the correct person.

Cover letter for a job application. One type of transmittal letter is the cover letter that accompanies a résumé and job application (Figure 10.3). The purpose of a job application cover letter is to explain how your credentials fit this particular job and to convey a sufficiently professional persona for the prospective employer to decide that you warrant an interview. Another purpose is to highlight some specific qualifications of skills. For example, you may have "C ++ programming" listed on your résumé under the category "Programming Languages." But for one particular job application, you may wish to accentuate this item. You could do so by stating in your cover letter:

> You will note on my résumé that I am experienced with C++ programming. In fact, I am not only a skilled C++ programmer, I have also taught evening programming classes at Metro Community College.

Make sure the first paragraph of your cover letter states the position you are applying for, especially if you are applying to the Human Resources Department. For example, you could begin with:

> Please consider my application for the senior software position advertised in this week's *Daily Chronicle*.

Inquiry letter. You may need to inquire about a product, service, set of specifications, or other item. As with most technical communication, keep your inquiry letter short and to the point. Follow the direct approach (see Figure 10.6 on page 198) and state clearly at the outset what you are requesting and why. Make sure you provide multiple ways for the recipient or the material to reach you: email, fax, phone, surface mail.

Word-processing templates. When discussing letters, it's important to note that most word-processing software (such as *Microsoft Word*) allows you to select from templates or predesigned letter formats. These templates usually provide fields for you to insert your name, your company name, and your message. Some templates provide background artwork or other decorative features. As tempting as it may be to simply choose a template, make sure the one you use is appropriate for your audience and purpose. If not, either use a blank document, or modify the template to suit your needs. Figure 10.4 shows a selection of typical word-processing templates.

Typical components of a letter

Most letters have the same basic components. Many organizations have set formats they follow for writing letters, so depending on where you work, these parts may appear in different locations on the page. But in general, a letter contains the elements listed on page 194 and following.

203 Elmwood Avenue
San Jose, CA 10462
April 22, 2000

Sara Costanza
Personnel Director
Liberty International, Inc.
Lansdowne, PA 24135

Dear Ms. Costanza:

Please consider my application for a junior management position
at your Lake Geneva resort. I will graduate from San Jose City
College on May 30 with an Associate of Arts degree in Hotel/
Restaurant Management. Dr. H. V. Garlid, my nutrition professor,
described his experience as a consultant for Liberty International
and encouraged me to apply.

For two years I worked as a part-time desk clerk, and I am now the
desk manager at a 200-unit resort. This experience, combined with
earlier customer relations work in a variety of situations, has given
me a clear and practical understanding of customers' needs and
expectations.

As an amateur chef, I know of the effort, attention, and patience
required to prepare fine food. Moreover, my skiing and sailing
background might be assets to your resort's recreation program.

I have confidence in my hospitality management skills. My
experience and education have prepared me to work well with
others and to respond creatively to changes, crises, and added
responsibilities.

Should my background meet your needs, please phone me any
weekday after 4 p.m. at 214-316-2419.

Sincerely,

James D. Purdy

James D. Purdy

Enclosure

FIGURE 10.3 ■ A sample cover letter for a job application.

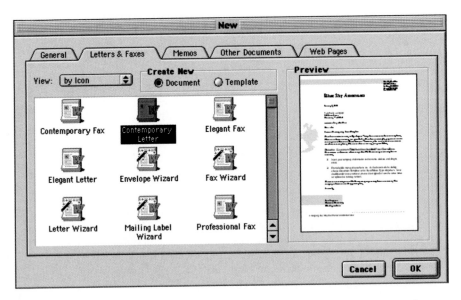

Figure 10.4 ■ **A letter in** *Microsoft Word.* You can choose from a variety of templates, such as the one highlighted above ("Contemporary Letter"). The clip art of planet Earth in the background of this template may not be appropriate for your situation, so don't choose a template if it's not right for your needs.
Source: Reprinted by permission from Microsoft Corporation.

Heading and date. If your stationery has a company letterhead, simply include the date two lines below the letterhead at the right or left margin. On blank stationery, include your return address and the date (but not your name):

> 154 Sea Lane
> Harwich, MA 02163
> July 15, 2000

Use the Postal Service's two-letter abbreviations (e.g., MA for Massachusetts; NY for New York; ND for North Dakota) in your heading, inside address, and on the envelope.

Inside address. Two line spaces (returns) after the heading and date information, flush against the left margin, is the inside address (the address to which you are sending the letter):

> Dr. Ann Mello, Dean of Students
> Western University
> 30 Mogul Hill Road
> Stowe, VT 52350

When possible, address a specific person and include his or her title.

Salutation. The salutation, two line spaces below the inside address, begins with "Dear" and ends with a colon ("Dear Ms. Smith:"). If you don't know the person's name, use the position title ("Dear Manager:"). Only address the recipient by first name if that is the way you would address this individual in person. Examples of salutations include

> Dear Ms. Martinez:
>
> Dear Managing Editor:
>
> Dear Professor Lee:

Remember not to use sexist language, such as "Dear Sir" or "Dear Madam." Instead, use the position title ("Dear Sales Manager").

Body text. Typically, your letter text begins two line spaces (returns) below the salutation. Workplace letters typically include

1. A brief introductory paragraph identifying you and your purpose
2. One or more body paragraphs containing the details of your message
3. A conclusion paragraph that sums up and encourages action

Keep the paragraphs short whenever possible. If your body section is too long, divide it into shorter paragraphs or place some items in a list (see Figure 10.5).

Complimentary closing. The closing, two line spaces (returns) below the last line of text, should parallel the level of formality used in the salutation and should reflect your relationship to the recipient (polite but not overly intimate). The following closings are listed in decreasing order of formality:

> Sincerely, (most often used)
>
> Respectfully,
>
> Cordially,
>
> Best wishes,
>
> Warmest wishes,
>
> Regards,
>
> Best,

Align the closing flush against the left margin.

LEVERETT LAND & TIMBER COMPANY, INC.   creative land use
                                        quality building materials
                                        architectural construction

January 17, 2001

Mr. Thomas E. Muffin
Clearwater Drive
Amherst, MA 01022

Dear Mr. Muffin:

I have examined the damage to your home caused by the ruptured water
pipe and consider the following repairs to be necessary and of immediate
concern:

> Exterior:
>   Remove plywood soffit panels beneath overhangs
>   Replace damaged insulation and plumbing
>   Remove all built-up ice within floor framing
>   Replace plywood panels and finish as required
>
> Northeast Bedroom—Lower Level:
>   Remove and replace all sheetrock, including closet
>   Remove and replace all door casings and baseboards
>   Remove and repair windowsill extensions and moldings
>   Remove and reinstall electric heaters
>   Respray ceilings and repaint all surfaces

This appraisal of damage repair does not include repairs and/or
replacements of carpets, tile work, or vinyl flooring. Also, this appraisal
assumes that the plywood subflooring on the main level has not been
severely damaged.

Leverett Land & Timber Company, Inc. proposes to furnish the necessary
materials and labor to perform the described damage repairs for the
amount of six thousand one hundred and eighty dollars ($6,180).

Sincerely,

P.A. Jackson

Gerald A. Jackson, President
GAJ/ob
Enc. Itemized estimate

FIGURE 10.5 ■ A sample business letter.

Signature. Type your full name and title on the fourth and fifth lines below the closing, aligned flush left. Sign your name in the triple space between the closing and your typed name.

Sincerely,

Meredith M. Crotin

Meredith M. Crotin
Principal Researcher

Specialized components of a letter

Some letters also have specialized parts, such as the following:

Attention line. Use an attention line when you write to an organization and do not know your recipient's name but are directing the letter to a specific department or position. Place the attention line flush with the left margin two line spaces (returns) below the inside address.

Glaxol Industries, Inc.
232 Rogaline Circle
Missoula, MT 61347

ATTENTION: Director of Research and Development

Subject line. Typically, subject lines are used with memos, but if the recipient is not expecting your letter, a subject line is a good way of catching a busy reader's attention.

SUBJECT: *New patent for hybrid wheat crop*

Place the subject line below the inside address or attention line. You can underline the subject to make it more prominent.

Typist's initials, distribution and enclosure notations. These items typically go at the bottom of the page. If someone else types your letter for you (common in the days of typewriters but rare today), your initials and your typist's initials appear as follows:

GLJ/pl

If you distribute copies of your letter to other recipients, indicate this by inserting the notation "cc." This notation once stood for "carbon copy," but no one uses carbon paper any more, so now it is said to stand for "courtesy copy."

cc: J. Hailey, S. Patel

Similarly, if you enclose any items, you can note this at the bottom of the page:

Enclosures: Certified checks (2), KBX plans (1)

Usability considerations

Organize for the situation. You can choose from two basic organizing patterns. The first is *direct,* in which you begin with the most important information (your request or the conclusion of your ideas) and then give the details supporting your case. The second is *indirect*, in which you start with the details and build toward a request or conclusion.

Figure 10.6 illustrates each approach. A direct approach is usually best, because readers can get right to the main point. But when you have to convey bad news or make a request that could be received unfavorably, it's often better to build your case first and then make your claim or request toward the end. When complaining about a faulty product, for example, a direct approach would probably be better, because the customer service person, who could easily receive hundreds of letters each day, will get to the point quickly. But for a letter requesting a raise in your consulting fee, an indirect approach is more appropriate.

Address the letter properly. A letter that is addressed to the wrong person will not be very effective. Before sending a letter, spend time researching the name of a person to whom you can send it.

Use language at an appropriate technical level. If your letter is too technical, recipients will not understand the material. But if your letter is too simplistic, you may insult or bore your reader. Learn all you can about your audience's level of expertise so you can choose appropriate language.

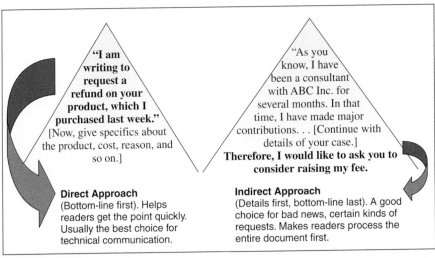

Figure 10.6 ■ Use a direct approach most of the time, but use an indirect approach when you have negative information to convey. If you do use the indirect approach, keep your information short and to the point, so your reader does not need to read a long narrative to get to the point.

Include all necessary information. If you are writing to ask for an updated version of your software, but you forget to include the current version number, the recipient will not be able to help. If you are lucky, that person will try to reach you and ask for more information. But quite possibly, your letter will be moved to the "get back to this" pile, and you may never receive a reply. Provide all the details your recipient will need.

Check spelling of proper nouns. Make sure you spell the name of the recipient, the company, and the product correctly. If the company is known as "Electronics, Inc." don't use "Electronic Corp." And, it may sound obvious, but be sure you've spelled your own name correctly, too!

Clearly state the main point. Whether you use a direct or indirect organizational pattern, state your main point clearly and directly. If you want a refund on a product, *make sure* you say, "I am writing to request a refund" or something equally direct.

Avoid puffed up language. Avoid the stuffy, puffed up, tired phrases some writers think they need to make their letters sound important. Here are a few of the old standards with some clearer, more direct translations:

Avoid	*Use instead*
As per your request	As you requested
Contingent upon receipt of	As soon as we receive
Please be advised that I	I
In accordance with your request	As you requested
Due to the fact that	Because

In the legal profession (and others), certain phrases such as these are known as "terms of art" and connote a specific concept. In these cases, you may not be able to avoid such phrases. But otherwise, strive for a simple, clear style.

Short reports

The situation

Reports present ideas and facts to interested parties, decision makers, and other audiences. Technical professionals rely on short reports as a basis for informed decisions on matters as diverse as the most comfortable office chairs to buy or the best recruit to hire for management training. Unlike long reports, short ones (2–5 pages) do not contain a lot of detail. For example, a long report describing the geologic conditions of an area might include an appendix with detailed comparisons of topsoil, ground water, and other conditions. A short report would summarize this information in a brief table, or, depending on the audience's prior knowledge, omit this information altogether.

Short reports are appropriate in a variety of situations. When the purpose of your communication is to inform an audience, offer a solution to a problem, report progress, or make a recommendation, you may wish to use a short report. Short reports often use a memo-like structure, starting with a memo-style header and breaking the text up into chunks separated by headings, but they contain more information than the typical memo.

Audience and purpose analysis

Do your best to determine who will read this report. For instance, even if the report is addressed to team members, it may be sent on to other managers, the legal department, or sales and marketing. If you can learn about the actual audience members in advance, you can anticipate their needs as you create the report. Also, before you start the report, be clear about its true purpose. For example, you may be under the impression that the report is intended to simply inform your colleagues about the new technical specifications for a component part. But after you interview some audience members and begin researching the content, you learn that what your audience really wants is a report that makes recommendations. Should your company invest in these new parts, or should they continue with the parts they currently use? Recommending is different from informing. It requires you to weigh the evidence, examine the data, and decide what you think is right.

Types of short reports

Short reports come in many types, depending on the situation. Common types include:

Recommendations. Recommendation reports interpret data, draw conclusions, and make recommendations, often in response to a specific request. The recommendation report in Figure 10.7 is addressed to the writer's boss. This is just one example of a short report used to examine a problem and recommend a solution.

Progress. Many organizations depend on progress reports (also called status reports) to track activities, issues, and progress on various projects. Some professions require regular progress reports (daily, weekly, monthly), while others may use these documents on an ad hoc basis, as needed to explain a specific project or task. Managers often use progress reports to evaluate projects and decide how to allocate funds. Figure 10.8 shows a progress report.

Meeting minutes. Many team or project meetings require someone to record the proceedings. Minutes are the records of such meetings. Copies of these minutes are usually distributed to all team members and interested parties. Often, organizations have templates or special formats for recording minutes. Meeting minutes are often distributed via email.

Typical components of short reports

Cover memo style heading. Many short reports begin with a brief cover memo explaining what the report contains.

Headings for major sections. Use headings for the major sections. Headings allow readers to quickly scan the report before launching into the text itself, and they provide a road map for the entire document.

Body text. Use a standard font and keep body text brief and to the point.

Bulleted lists and visuals. Use numbered or bulleted lists if the content warrants it. Lists help readers to skim the document and help to order the information. But too many lists can be confusing, so use lists sparingly to make certain key points visible.

Usability considerations

Use effective page layout and document design. The longer a document, the more navigation tools you should include to help readers find what they need and stay focused on the material. Even a short (2–3 page) report of solid, unbroken text (no font changes, no headings) can quickly become hard to read. Use headings, numbered lists, page numbers, and font changes to increase readability.

Perform your best research. Make sure you've gathered the right information before you begin writing the report. For the recommendation report in Figure 10.7, the writer did enough research to rule out one problem (computer display screens as the cause of employee headaches) before settling on the actual problem (excessive glare on display screens from background lighting). Research may include interviewing people, using the library, searching for information on the Internet, and taking informal surveys (Chapter 4).

Use visuals as appropriate. Insert visuals as needed. In a short report, you don't want to bog readers down with excessive graphic information, but a few well-chosen graphs, tables, or charts can help clarify your point (Chapter 9).

Address the purpose. Remember the purpose when you create the report. If your purpose is to recommend, don't take a neutral stance and simply state the

Trans Globe Airlines

MEMORANDUM

To: R. Ames, Vice President, Personnel
From: B. Doakes, Health and Safety
Date: August 15, 20xx
Subject: **Recommendations for Reducing Agents' Discomfort**

In our July 20 staff meeting, we discussed physical discomfort among reservation and booking agents, who spend eight hours daily at automated workstations. Our agents complain of headaches, eyestrain and irritation, blurred or double vision, backaches, and stiff joints. This report outlines the apparent causes and recommends ways of reducing discomfort.

Causes of Agents' Discomfort

For the time being, I have ruled out the computer display screens as a cause of headaches and eye problems for the following reasons:

1. Our new display screens have excellent contrast and no flicker.
2. Research findings about the effects of low-level radiation from computer screens are inconclusive.

The headaches and eye problems seem to be caused by the excessive glare on display screens from background lighting.

Other discomforts, such as backaches and stiffness, apparently result from the agents' sitting in one position for up to two hours between breaks.

Recommended Changes

We can eliminate much discomfort by improving background lighting, workstation conditions, and work routines and habits.

Background Lighting. To reduce the glare on display screens, these are recommended changes in background lighting:

1. Decrease all overhead lighting by installing lower-wattage bulbs.
2. Keep all curtains and adjustable blinds on the south and west windows at least half-drawn, to block direct sunlight.
3. Install shades to direct the overhead light straight downward, so that it is not reflected by the screens.

FiGURE 10.7 ■ A sample recommendation report in memo format.

Workstation Conditions. These are recommended changes in the workstations:

1. Reposition all screens so light sources are neither at front nor back.
2. Wash the surface of each screen weekly.
3. Adjust each screen so the top is slightly below the operator's eye level.
4. Adjust all keyboards so they are 27 inches from the floor.
5. Replace all fixed chairs with adjustable, armless, secretarial chairs.

Work Routines and Habits. These are recommended changes in agents' work routines and habits:

1. Allow frequent rest periods (10 minutes each hour instead of 30 minutes twice daily).
2. Provide yearly eye exams for all terminal operators, as part of our routine health-care program.
3. Train employees to adjust screen contrast and brightness whenever the background lighting changes.
4. Offer workshops on improving posture.

These changes will give us time to consider more complex options such as installing hoods and antiglare filters on terminal screens, replacing fluorescent lighting with incandescent, covering surfaces with nonglare paint, or other disruptive procedures.

cc. J. Bush, Medical Director
M. White, Manager of Physical Plant

Figure 10.7 ■ *(Continued)*

Subject: **Progress Report: Equipment for New Operations Building**

Work Completed

Our training group has met twice since our May 12 report in order to answer the questions you posed in your May 16 memo. In our first meeting, we identified the types of training we anticipate.

Types of Training Anticipated

- Divisional Surveys
- Loan Officer Work Experience
- Divisional Systems Training
- Divisional Clerical Training (Continuing)
- Divisional Clerical Training (New Employees)
- Divisional Management Training (Seminars)
- Special/New Equipment Training

In our second meeting, we considered various areas for the training room.

Training Room

The frequency of training necessitates having a training room available daily. The large training room in the Corporate Education area (10th floor) would be ideal. Before submitting our next report, we need your confirmation that this room can be assigned to us.

To support the training programs, we purchased this equipment:

- Audioviewer
- LCD monitor
- Videocassette recorder and monitor
- CRT
- Software for computer-assisted instruction
- Slide projector
- Tape recorder

This equipment will allow us to administer training in a variety of modes, ranging from programmed and learner-controlled instruction to group seminars and workshops.

Figure 10.8 ■ Sample progress report in memo format.

Work Remaining

To support the training, we need to furnish the room appropriately. Because the types of training will vary, the furniture should provide a flexible environment. Outlined here are our anticipated furnishing needs.

- Tables and chairs that can be set up in many configurations. These would allow for individual or group training and large seminars.
- Portable room dividers. These would provide study space for training with programmed instruction, and allow for simultaneous training.
- Built-in storage space for audiovisual equipment and training supplies. Ideally, this storage space should be multipurpose, providing work or display surfaces.
- A flexible lighting system, important for audiovisual presentations and individualized study.
- Independent temperature control, to ensure that the training room remains comfortable regardless of group size and equipment used.

The project is on schedule. As soon as we receive your approval of these specifications, we will proceed to the next step: sending out bids for room dividers, and having plans drawn for the built-in storage space.

cc.   R. S. Pike, SVP
      G. T. Bailey, SVP

FIGURE 10.8 ■ *(Continued)*

facts. If your purpose is to show progress, make it clear at the outset what progress has been made. Use the direct approach (Figure 10.6). For example:

> Our goals for this month were to finish the Alpha project and begin planning the Beta project. We have exceeded these goals, and this report will describe our current progress.

Use appropriate headings. Use headings that make sense to your readers. The memo in Figure 10.7 is essentially an overview of a problem followed by a recommended solution, and the headings ("Cause of agents' discomfort" "Recommended changes") make this clear.

Write clearly and concisely. Get to the point quickly. Use a brief introductory paragraph that states your case or conclusion. Don't bog your reader down with unnecessary background or history sections.

ORAL COMMUNICATION

The situation

In addition to being good writers and designers, technical communicators need to present their ideas effectively in person. The skills required include stage presence, excellent research, and strong organization. You may be asked to make oral presentations at meetings in your department or company (status reports, team meetings, procedure updates), at professional conferences or meetings, to your community, or in the classroom.

Unlike writing, oral presentations are truly interactive. Face-to-face communication is arguably the richest form, because you can give and receive information using body language, vocal inflection, eye contact, and other physical features. In addition, there is room for give and take, which does not happen with traditional written documents. Oral presentations allow you to see how your audience reacts. You can get immediate feedback, and you can change or amend your ideas on the spot.

Most professionals use presentation software such as *Microsoft PowerPoint* when giving a presentation. But despite the colors and easy-to-use templates such products offer, you are still responsible for a presentation that is well researched and professionally delivered.

Audience and purpose analysis

Learn about an audience's attitudes and biases toward and personal experiences with your subject, and see if you can find out exactly who will be attending and what their role is within the organization. A person's role in the organization will affect how they listen to your topic: Managers may care about the bottom line, while designers may care more about how much time they will have to work on the interface. Oral presentations also require you to consider the feelings of the group as a whole. When you are speaking to a live audience, people's attitudes and ideas can rub off on others. If one person raises an issue with your topic, others might be reminded that they, too, are interested in this issue. The next thing you know, your audience will become a group, not a set of individuals.

In defining the purpose of your presentation, understand how your audience will use the information you are presenting. Do they need this information to perform tasks? What do they need to know immediately? How can you make the best use of this actual time with them, and should you save certain items for email, memos, or Web discussions? Is your purpose to inform your audience, persuade them, or train them? Or is your presentation part of a larger context, such as a conference presentation? Customize your audience and purpose analysis to address these questions, which are specific to oral presentations.

Types of oral presentations

Many types of oral presentations are common in science, business, and technical communication, including the following:

Informative presentations. When your audience needs factual information about products, procedures, technical topics, or other items, give an informative presentation. Informative presentations are often given at conferences, product update meetings, briefings, or class lectures. In an informative presentation, your goal is to be as impartial as possible and to provide the best information you can locate.

Training sessions. Training sessions teach audience members how to perform a specific task or set of tasks. Training can cover areas such as on-the-job safety, how to use a specific software application, or how to exit a capsized kayak. Some technical communicators specialize in training.

Persuasive presentations. Persuasive presentations are designed to change an audience's opinions. For example, an engineer at a nuclear power plant may wish to persuade her peers that a standard procedure is unsafe and should be changed. In a persuasive situation, you need to perform adequate research so that you are well informed on all sides of the issue.

Action plans. A more specific form of persuasive presentations, action plans are appropriate in situations where you want your audience to take a particular action. If you not only wanted to convince other engineers about a design flaw, but also wanted the company to take specific action, you would give an action plan presentation that outlined the problem, presented a specific solution, and then tried to convince your audience to take the sort of action needed to implement the solution.

Sales presentations. Sales presentations blend informative and persuasive elements. Usually, the speaker presents information about a product or service in a way that will persuade the audience to buy it. Technical sales presentations need to be well researched: At many high-tech companies, technical sales representatives are often scientists or engineers who understand the product's complexities but also have a knack for effective communication.

In addition to presenting in front of live audiences, you may be asked to present via interactive television, satellite, or a live connection via the Internet. The more technology involved, the more complicated the presentation, so be sure you have technical experts to assist you. In particular, make sure your overhead slides or computer presentations will show up in these electronic formats.

You may also be asked to give a presentation in another person's absence; for example, a speaker is suddenly ill, or her flight gets snowbound in the Chicago airport. In these cases, get as much information as possible from the original speaker.

Typical components of oral presentations

Introduction. The introduction to a presentation is your chance to set the stage. For most presentations, you have three main tasks:

1. Capture your audience's attention by telling a quick story, asking a question, or relating your topic to a current event or something else the audience cares about.
2. Establish your credibility by stating your credentials or explaining where you obtained your information.
3. Preview your presentation by listing the main points and the overall conclusion.

An introduction following this format might sound something like this:

> How many times have you searched for medical information on the Web, only to find that after hours online, you can't locate anything useful? If you're like most Americans polled in a recent survey, you may feel that you are wasting time when it comes to Web-based medical information. My name is Travis Armstrong, and I've been researching this topic for a term paper. Today, I'd like to share my findings with you by covering three main points: how to search for medical information, how to separate good information from bad, and how to contribute to medical discussions online.

Body. Readers who get confused or want to know the scope of a written document (reports, manuals, and so on) can look back at the headings, table of contents, or previous pages. But oral presentations do not have these features.

Therefore, you must give your audience a well-organized presentation that is interesting and easy to follow. Structure the material in small chunks. To signal that you are moving from one main point to another, use transition statements, such as, "Now that I've explained how to separate good information from bad, let me tell you how you can contribute to medical discussions on the Internet."

Conclusion. Your conclusion should return full circle to your introduction. Remind your audience of the big picture, restate the points you've just covered, and leave listeners with some final advice or tips for locating more information. You can also distribute handouts during this time.

Computer projection software and other visual aids. Most presentations use visuals of some sort: overhead transparencies, flip charts, or computer projection software. Keep in mind that the more technology you use, the more prepared you must be. Used properly, computer software can make presentations interesting and enjoyable, because they allow the use of color, graphics, clip art, and images from the Web (see Figure 10.9).

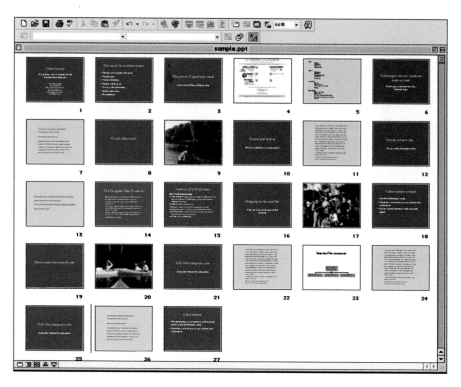

FɪɢuʀE **10.9** ■ **Sample slides from a *PowerPoint* presentation.** Don't let the software trick you into thinking you don't need to practice or do research.
Source: PowerPoint screen shot is reprinted by permission from Microsoft Corporation.

Handouts. If you bring handouts to a presentation, consider when you want to give them to the audience. The moment you provide handouts, your audience will look at the handouts and not at you.

Time constraints. Most presentations have a time limit. If you are presenting at a conference on a panel of three or four people, you may be asked in advance to limit your talk to 15 minutes. If you are presenting at a one-hour meeting, you should consider the amount of time people are able to concentrate and remember to save time for questions. Exceeding your allotted time (usually no more than 20 minutes in most business settings) is inconsiderate of your audience and the other speakers. Practice your presentation and make it fit the time allowed.

Usability considerations

Maintain confidence and project a professional persona. Fear of public speaking is very common, even for the most skilled speakers. The best way to avoid nervousness is to perform solid research on the topic, prepare well, and practice. Deep breathing and positive thinking also help with unsteady nerves. The more confident you appear, the more credible your message.

Be ready to adapt to your audience. Speaking to a live audience means you may need to adapt on the spot. If you start your talk but quickly realize that your audience does not understand some of the technical terms, you need to be ready to stop, explain some of the terms in simpler language, and perhaps even reconsider whether to do certain parts of your presentation at all. The more experience you gain with giving presentations, the better you will become at learning to "read" your audience.

Prepare for your technology to fail. No matter how many disk copies you brought, how fast the Internet connection, or how great the technical support staff, always be prepared for your technology to fail. Computer presentation software is dynamic and professional, but it does not always work. Overhead projectors are excellent for presenting outlines of your talk, visuals, and other material, but what do you do if the bulb burns out? Always have a backup plan: Paper handouts (and a spare bulb) are a good idea.

Get a sense of the physical layout of the room. Make sure you know the size and layout of the room before you show up. Is it a conference room or a large lecture hall? The type size chosen for slides to be shown in a conference room may not be large enough for a lecture hall. If you can, visit the space in advance.

Spell check your slides. Even though most people spell check documents, for some reason they often forget to spell check their presentation slides. You don't want the audience distracted by misspellings on the screen when you are speaking.

Use your memory and practice your delivery. Don't read your presentation from notecards or from a prepared speech. Instead, seek a style that is professional but natural. Memorize the key phrases or concepts in your presentation, and use these key terms to jog your memory about the other items you wish to mention. Figure 10.10 shows an outline for an oral presentation. Use bullet points on your overheads or slides to help guide you. Practice until you can speak with confidence. You should practice in front of friends or sample audience members and ask for their feedback. Or, tape or video yourself and use this to help improve your style.

Medical Information on the Web
Oral Presentation by Travis Armstrong

Outline

Introduction
    Ask questions (How many times have you searched . . .)
    Introduce myself
    Outline main points
        Searching
        Separating good from bad
        Contributing to discussions online

Body
    Searching for medical information
    Separating good from bad information
    Contributing to discussions online

Conclusion
    Restate my main points
    Give sources for more information and handout

Figure 10.10 ■ **Sample outline for a presentation.** Make sure you practice and know your subject well. Use the main points to guide you along.

Review Checklist

Type of Communication	Use When You Need to	Specific Types
Email	transmit information quickly and efficiently, often attaching formatted documents or Web addresses	*Microsoft Outlook, Netscape Mail,* or Web sites that let you read email
Memos	transmit information that may be longer than you would transmit by email and less formal than by letter	transmittal memo, meeting minutes, brief report
Letters	transmit information that may be longer and more formal than you would transmit by email	transmittal letters, cover letters, inquiry letters
Short reports	present ideas and facts to interested parties, decision makers, and other audiences	recommendations, progress reports, meeting minutes
Oral communication	present at meetings in your department or company, at professional conferences, or in the classroom	informative, training, action plans, sales presentations

Exercises

The Web site for this textbook (www.ablongman.com/gurak) contains links to sample documents online as well as exercises related to different document types.

1. **Focus on ▶Writing.** People regularly contact your organization (your company, agency, or college department) via email, letter, or your Web site to request information. To answer these many inquiries, you decide to prepare a frequently asked questions (FAQ) list in response to the ten most asked questions about products, services, specific concentrations within

the major, admission requirements, or the like. In addition to being posted on your Web site, this list can be sent as an email attachment or mailed out as hard copy, depending on your reader's preference.

After analyzing your specific audience and purpose and doing the research, prepare your list in short report format, paying careful attention to the usability considerations in Chapter 3.

2. **Focus on ⮕Writing.** Your boss or college dean wants recommendations about whether to begin electronic monitoring of email correspondence and Internet use at your company or your school. A few of your audience's major questions: What are the pros and cons? What are the ethical considerations? Should we do it routinely? Should we do it selectively? Should we rule it out? Is more inquiry needed? Should we allow the entire organization to share in the decision?

Do the research and prepare a recommendation memo that makes your case reasonably and persuasively (review Chapter 6 for ethical considerations).

3. Write an actual letter to a business about one of their products following the format described in this chapter. Exchange letters with another student and check each other's letters using the "rules" from this chapter. Rewrite and send the letter; if you get a response, share it with the class.

4. Make up a scenario related to your career/future career about an oral presentation you may have to make. Make sure you give all related background information about your job. For example: You are a health care retailer who sells medical equipment. You need to present a new type of clamp designed by your company that you hope the hospital administrators and doctors will want to purchase at the workshop they are attending. Describe what type of oral presentation you would give, do an audience/purpose analysis, and tell what you would use to enhance your presentation. Compare your scenario with those of other students in your class.

The Collaboration Window

Writing does not take place in isolation. Even if you are the main author of a memo or report, you will work with others to gather information, analyze your findings, learn about user tasks, and more. With your profession/major in mind, choose a form of communication described in this chapter (letter, report, etc.) and describe how you will need to work with others in your field when composing the document. Write your thoughts in a memo or email to your professor.

The Global Window

Interview someone whose work takes them to one or more countries outside the United States, and ask that person to describe for you which document types are used in which situations and in different countries and settings. For example, is a paper letter always considered formal? Is letterhead paper the same size across the world? In the United States, email seems to be replacing the paper memo. Is this true in other countries as well? If you have trouble locating someone to interview, try asking your instructor for the name of someone who hires interns from your program. Or, you could ask your instructor for ideas about faculty members who travel abroad.

Click on This

Technical communicators often discover that after they have prepared a short report for distribution on paper, someone in the organization says, "Hey, let's put this up on the Web, too!" But paper documents may need to be rewritten and restructured before they will be usable as Web documents. To learn more about creating documents for a Web environment, visit the HTML Writer's Guild, "the world's largest international organization of Web authors with over 110,000 members in more than 150 nations worldwide," at www.hwg.org.

Product-Oriented Communication Situations

Specifications

Brief instructions

Procedures

Documentation and manuals

Technical marketing material

Review checklist

Exercises

The collaboration window

The global window

Click on this

Specifications

The situation

Airplanes, bridges, computer software, and countless other technologies are produced according to specifications. A particularly exacting type of description, specifications (or "specs") prescribe standards for performance, safety, and quality. For virtually any product, specifications describe features such as methods for manufacturing, building, or installing a product; materials and equipment used; and size, shape, and weight. Specifications are often used to ensure compliance with a particular safety code, engineering standard, or government or other ruling. Because specifications define an "acceptable" level of quality, any product that fails to meet these specs may provide grounds for a lawsuit. When injury or death results (as in a bridge collapse or an airline accident), the device is usually checked to be sure it was built and maintained according to the appropriate specifications. If not, the contractor, manufacturer, or supplier may be liable.

Specifications are called for in situations where a technology or procedure must be executed in a precise manner. For example, the Institute of Electrical and Electronics Engineers (IEEE) issues a series of specifications (called "standards") that prescribe the technical format and specs for a range of devices, including circuits, electrical insulation, and telecommunications. Engineers are obligated to follow these standards if they want their products to be safe, as well as compatible with others in that field. Specifications are also important in situations where many professionals from different backgrounds work together on a project. These specs help ensure that contractors, architects, landscapers, and others have a master plan and use the same parts, materials, and designs. Software developers also follow specifications when creating computer programs.

Audience and purpose analysis

Specifications may be written for a wide range of readers, including customers, designers, contractors, suppliers, engineers, programmers, and inspectors. An audience analysis will help you determine who will be reading the specs. If your audience consists primarily of technical experts (such as engineers who use the IEEE standards), you can use specialized language and succinct explanations (see Figure 11.1). But if your audience is a mixed group, you may need to include more detail, or you may need to refer readers to other sources of information (a glossary, Web site, or an attachment).

In terms of purpose, specifications are useful when your audience needs to understand and agree on what is to be done and how it is to be done. In addition to guiding how a product is designed and constructed, specifications can also help people use and maintain a product. For instance, specifications for a

IEEE C62.1-1989 - Description

IEEE
*Networking
the World*™

IEEE C62.1-1989 - revision of ANSI/IEEE C62.1-1984
IEEE Standard for Gapped Silicon-Carbide Surge Arresters for AC Power Circuits
Abstract: IEEE C62.1-1989, *IEEE Standard for Gapped Silicon-Carbide Surge Arresters for AC Power Systems,* describes the service conditions, classifications and voltage ratings, design tests with corresponding performance characteristics, conformance tests, and certification test procedures for station, intermediate, distribution and secondary class arresters. Terminal connections, housing leakage distance, mounting and identification requirements are defined. Definitions are provided to clarify the required test procedures and other portions of the text.
Contents

1. Scope

2. Definitions

3. References

4. Service Conditions
   4.1   Usual Service Conditions
   4.2   Unusual Service Conditions

5. Classification and Voltage Rating of Arresters
   5.1   Voltage Ratings
   5.2   Test Requirements

Voltage Withstand Tests
Power-Frequency Sparkover Test

Fɪɢᴜʀᴇ 11.1 ■ **A sample from an IEEE standard for surge protectors.**

color inkjet printer (see Figure 11.2) include the product's power requirements, noise emissions, and weight and size of paper. Product support literature for appliances, power tools, and other items often contain specifications
so users can select an appropriate operating environment or replace worn
parts. Specifications can also be important in technical marketing material
(discussed later in this chapter) by helping potential customers see the details
of the product.

Types of specifications

Industry standards. Many industries issue specifications and standards for
products in the field. The IEEE, for example, issues standards for electrical and
electronics devices.

Government standards. Government organizations, such as the Food and Drug
Administration (FDA), the Consumer Product Safety Commission, and others,
produce guidelines and standards for a multitude of items. The FDA, for example, regulates the production and sale of food products and vitamin supplements.

Functional specs. Functional specs are used in the private sector to outline exactly what needs to be done on a project (Figure 11.3). A functional spec does
not always indicate how certain parts of the project will be implemented, however. For example, a functional spec for a software product may indicate that
the software should allow users to log in, enter data, and move between
screens. Software development teams use this functional spec to guide them as
they write and test the computer code.

Internet specs. The World Wide Web consortium (www.w3c.org) is an indus-
try/university organization that develops common protocols (technical standards) for Web page design and development. These protocols provide standards for Web development in areas such as graphics, hypertext markup
language (HTML), and other technical aspects that allow the Web to function
across a variety of platforms, countries, and browsers.

Typical components of specifications

As noted, there are many types of specifications. The components of the document will often be dictated by the industry or organization. The IEEE surge
protector standard (Figure 11.1) follows the format for all similar IEEE documents. In general, specifications include:

A brief introduction or description. Most specs include some kind of overview
section, be it a one or two sentence introduction, an abstract, or a similarly

 Specifications

Power Requirements

Power Adapter (universal input)

Input Voltage: 100 to 240 VAC (±10%)
Input Frequency: 50 to 60 Hz (±3 Hz)

Automatically accommodates the world-wide range of AC line voltages and frequencies. There is no on/off switch on the power adapter.

Declared noise emissions in accordance with ISO 9296:

Sound power level. LWAd (1B=10dB): 5.5 B in Normal mode.

Sound pressure level. LpAm (bystander positions): 42 dB in Normal mode.

Media Weight

Paper:	16 to 110 lb index (60 to 200 gsm)
Envelopes:	20 to 24 lb (75 to 90 gsm)
Cards:	110 lb index max; 0.012 in max thickness (110 to 200 gsm; 0.3 mm max thickness)
Banner Paper:	20 lb (75 gsm)

Media Handling

Sheets:	up to 100 sheets
Banners:	up to 20 sheets
Envelopes:	up to 15 envelopes
Cards:	up to 30 cards
Transparencies:	up to 25 sheets
Labels:	up to 20 sheets of Avery paper labels
	Use only U.S. letter-sized or A4-sized sheets. Use only paper labels specifically designed for use with HP inkjet printers.
OUT tray capacity:	up to 50 sheets

Media Size

Custom size:	
Width:	3.0 to 8.5 in (77 to 216 mm)
Length:	3 to 14 in (77 to 356 mm)
U.S. letter:	8.5 x 11 in (216 x 279 mm)
Banner	
U.S. letter:	8.5 x 11 in (216 x 279 mm)
U.S. legal:	8.5 x 14 in (216 x 356 mm)
Executive:	7.25 x 10.5 in (184 x 267 mm)
U.S. No. 10 envelope:	4.13 x 9.5 in (105 x 241 mm)
Invitation A2 envelope:	4.37 x 5.75 in
Index card:	3 x 5 in (76 x 127 mm)
Index card:	4 x 6 in (102 x 152 mm)
Index card:	5 x 8 in (127 x 203 mm)
European A4:	210 x 297 mm
European A5:	148 x 210 mm
Banner	
European A4:	210 x 297 mm
B5-JIS:	182 x 257 mm
European DL envelope:	220 x 110 mm
European C6 envelope:	114 x 162 mm
European A6 card:	105 x 148 mm
Japanese Hagaki postcard:	100 x 148 mm

Figure 11.2 ■ **Specifications for a color inkjet printer.**

Source: Hewlett-Packard Company. Reprinted with permission.

Ruger, Filstone, and Grant
Architects

SPECIFICATIONS FOR THE POWNAL CLINIC BUILDING

Foundation
> footings: 8" x 16" concrete (load-bearing capacity: 3,000 lbs. per sq. in.)
> frost walls: 8" x 4' @ 3,000 psi
> slab: 4" @ 3,000 psi, reinforced with wire mesh over vapor barrier

Exterior Walls
> frame: eastern pine #2 timber frame with exterior partitions set inside posts
> exterior partitions: 2" x 4" kiln-dried spruce set at 16" on center
> sheathing: 1/4" exterior-grade plywood
> siding: #1 red cedar with a 1/2" x 6' bevel
> trim: finished pine boards ranging from 1" x 4" to 1" x 10"
> painting: 2 coats of Clear Wood Finish on siding; trim primed and finished
> with one coat of bone white, oil base paint

Roof system
> framing: 2" x 12" kiln-dried spruce set at 24" on center
> sheathing: 5/8" exterior-grade plywood
> finish: 240 Celotex 20-year fiberglass shingles over #15 impregnated felt
> roofing paper
> flashing: copper

Windows
> Anderson casement and fixed-over-awning models, with white exterior
> cladding, insulating glass and screens, and wood interior frames

Landscape
> driveway: gravel base, with 3" traprock surface
> walks: timber defined, with traprock surface
> cleared areas: to be rough graded and covered with wood chips
> plantings: 10 assorted lawn plants along the road side of the building

FIGURE 11.3 ■ Specifications for a building project.

brief description of the document. If your audience is completely familiar with the material, a title can serve this purpose.

List of component parts or materials. Specifications that deal with devices or hardware often list the individual parts. For example, if you purchase a new power tool, you will probably find a list of parts somewhere in the instruction manual.

Reference to other documents or specs. Often, one set of specifications refers to another. For example, a functional spec for a software product may refer readers to an earlier document. If you put your spec on a Web site, you can easily create links from the new document to the old one.

Usability considerations

Understand how people will use the document. Will your audience use the specs to build something (a bridge, a software product)? If so, they will need detailed information. But what if they need to refer to the specs only on occasion, say, if a part is needed? In this case, a spec can be shorter, for example, a list of parts with order numbers and brief descriptions.

Use the same terms to refer to the same parts or steps. If the specs list indicates "ergonomic adapter" in one section but then uses the term "iMac Mouse Adapter" in a later section, users may become frustrated and confused.

Use adequate retrieval aids. Especially in longer documents, some audience members may only be interested in one aspect of the spec. For example, a programmer working on a subset of the entire project may want to look up the technical details for her part of the project. If you use clear headings and a table of contents, people can find the parts they need.

Follow a standard format. If the organization follows a standard format for specs, use it. If not, you may want to create a template and suggest that the entire company use it. Standard formats help users find what they need, because each document created with a template will contain similar sections and subsections.

Keep it simple. People look at specs because they want quick access to items, parts, technical protocols, and so on. These users are not interested in reading a novel. If you can, keep your specs limited to short lists, using prose only as necessary.

Check your use of technical terms. Make sure your technical terms are standard for the industry. Be sure you have spelled these terms correctly or used the standard abbreviations. Remember that your spell checker won't find some technical terms, so you may wish to create a new dictionary within your word-processing software just for this project.

When giving instructions, use active voice. If the specs require you to instruct your audience, use active voice and imperative mood. For example, "Insert the bolt" or "Remove the wiring insulation."

Brief instructions

The situation

Surrounded as we are by technology, we are quite naturally also surrounded by instructions. Brief, to-the-point instructions are called for when you want to provide users with information on how to assemble, connect, and/or use a product. Even a product as simple as a do-it-yourself bookshelf requires clear, simple instructions. Brief instructions often appear as part of a larger set of instructions and procedures to help users get started immediately. Figure 11.4 is an example—it lists three steps users need to connect their drives to their computers. Users can refer to the card to get started immediately, and they can read the more extensive User's Guide later.

Audience and purpose analysis

Consider how much experience your audience has with the technology or task. If someone already owns a Zip drive at work, for example, and has purchased a new one for home, that person will need only a brief reminder of how to set up and operate the device. Novice users, on the other hand (users who are new to this technology), will need more information.

In terms of purpose, instructions are almost always written to help users perform a task or series of tasks. During your analysis, you need to discover what these tasks are and in what order people should perform them. For more about task-oriented communication, see Chapter 3.

Types of brief instructions

Quick reference cards. Figure 11.4 shows a quick reference card. The instructions fit on the single side of a card and contain only the basic steps necessary for the task. This card is designed to be used during an installation. Other

iomega. USB drive
Quick Install

zip. 100

@ *If you need more detailed install instructions, see the User's Guide.*

1 **Make sure the operating system has fully loaded, then install the IomegaWare™ software.**

2 **Connect the power supply.**

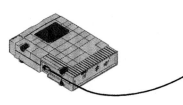

3 **Connect the Zip® USB cable to the Zip drive.**

NOTE: The drive can be connected while the computer is on.

4 **Connect the other end of the USB cable to the computer.**

NOTE: You can use either USB port.

CAUTION! **Do not use USB extension cables with your Zip USB drive. Data loss may result.**

Congratulations! Insert a Zip disk and your Zip USB drive is now ready to use.

iomega.

03656800 10/08/98 r19

FIGURE 11.4 ■ A quick install card. This card lists the three steps users need to perform in order to connect their new drives to their computers. Note the use of action verbs ("connect") and the smart use of diagrams.
Source: Reprinted with permission of Iomega Corporation.

cards, such as the kind you carry in your wallet with bank ATM instructions, are designed to be used over and over as needed.

Assembly instructions. These are the instruction sheets that come with most do-it-yourself devices or products. Assembly instructions usually contain numbered steps, a parts list, and diagrams.

Wordless instructions. In a global marketplace, it's important to have instructions that can be understood across language barriers. Translating instructions into dozens of languages is expensive and leads to bulky, cumbersome materials. Many companies have turned to wordless instructions, which use diagrams and arrows to explain a procedure (see Figure 11.5).

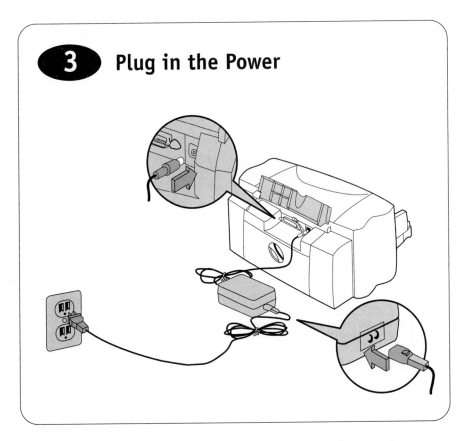

3 Plug in the Power

Figure 11.5 ■ **Wordless instructions.** This example uses a few words but primarily relies on an exploded diagram to explain the procedure.
Source: Hewlett-Packard Company. Reproduced with permission.

Typical components of instructions

Title. Most instructions, brief or extended, have a short title, such as "Quick Set Up Instructions for the Printer." Keep the title brief and to the point.

Quick overview of the task. In cases where the instructions will take more than a few steps, you can create a quick overview of the entire process, such as:

> To use your new SCSI adapter cable, you will first need to install the software, and then connect the cable.

Your instructions would then be divided into two parts: installing the software and connecting the cable. Overviews help users see the big picture before they embark on the detailed actions.

Step-by-step instructions. Most instructions are written in a numbered list format so users know the sequence of the tasks and can remember where they left off.

Diagrams. Diagrams can be extremely helpful, because they allow users to see the process or the parts to be assembled. Diagrams are especially important when you are creating a document for an international audience.

Follow-up information. It is a good idea to provide an address, phone number, Web site, and email address where users can obtain assistance.

Usability considerations

Define the task your users need to perform. Experts at writing instructions always perform a task analysis, which focuses on tasks users will need to perform in particular situations. For example, the main tasks on users' minds when they open the box for a new computer component, such as a printer, are how to set it up and use it.

Determine the size of the final document. Brief instructions are intended to help users perform tasks quickly. The final document should not be bulky or cumbersome. It should be easy to use and, if possible, printed on a single page. The quick reference card in Figure 11.4 is a good example.

Use the same terms to refer to the same parts or steps. Readers are frustrated by instructions that use terms inconsistently. For example, be sure your instructions don't say to connect the "SCSI printer cable" in one step but then refer to the "peripheral cable" in a later step.

Test your instructions. Most technical communication products should be proofread or tested by real audience members, but this usability concern is especially important with instructions. You won't know if the instructions work unless you watch a group of users try them out.

Use imperative voice and action verbs. Imperative sentences leave out the word "you" and begin with the verb. Use these sentences with strong action verbs when listing tasks. For example:

> Connect the cable.
> Insert the disk.
> Open the hatch.

See Chapter 3 and Appendix A for more information on this subject.

Keep it simple. Brief instructions should be simple and to the point. Include only the information that the user needs to perform that particular task. But direct the user to other sources (online help, Web site, manual) for further information if needed.

Procedures

The situation

Procedures are documents that provide information, steps, and guidelines for completing a task. Longer than brief instructions, some procedures are used to instruct and train employees. Others are used to meet legal requirements and ensure safety. For example, the U.S. Occupational Safety and Health Administration (OSHA) requires employees in certain workplaces (factories, construction sites, hospitals) to follow strict safety procedures. These procedures must be updated according to new laws and policies, and the written procedures must be available for employees to read.

Procedures are also useful in situations where you want to standardize a task. If your company has several employees all performing the same task but doing so with different computers, different software, or different styles, it may be necessary to standardize these procedures so that everyone's work is compatible. A written document, called a Standard Operating Procedure (SOP), becomes the formal explanation of how a particular task is done at that company.

Procedures are also part of the documentation and manuals that accompany new software, hardware, home appliances, and so on. Often, these procedures will be written as brief instructions. Other times, they are more elaborate, depending on the audience's technical background, experience, and needs.

Audience and purpose analysis

Determine what knowledge your readers already have about the procedure. In a manufacturing plant, for example, many workers may be extremely familiar with the tasks they perform on a daily basis. But if a team that normally works with extruding equipment is asked to take over on the mixing machines, those workers would need to be trained according to the standard procedures for that equipment. Your audience in this case would know the basics, because they've worked at the plant for years, but they would need detailed information on the specific machines, amounts of chemicals, and so on.

You may learn that the procedures will be used by multiple audiences, such as long-time employees, new trainees, and so on. In this case, you should consider using the *layered approach* (Chapter 3): a quick reference procedure for those who need to refresh their memories and a more detailed document for new workers.

In terms of purpose, you need to determine how your audience will use these procedures. Will they use them to assemble a new computer or install software? Will the procedures be used on the factory floor to remind workers of the safest way to perform a task? Can your audience access the procedures on a computer (online help or the Web), or do you need to create a paper document? Which medium is likely to be more effective in this situation? Once you know exactly how the procedures will be used, you can make decisions about the document's length, format, level of detail, and medium.

Types of procedures

Standard operating procedures (SOPs). Figure 11.6 illustrates a standard operating procedure for using a special microscope in video enhanced microscopy. This SOP is for students and researchers who wish to use the equipment. Because this procedure is lengthy, the document is broken into smaller chunks, easily accessible through a hyperlink (note that in a paper document, a table of contents would serve the same purpose). The procedures for each section are written in clear steps using imperative verbs such as "turn on," "place," and "focus."

Instructions (long). Some procedures are written in the form of long instructions. For situations where your audience needs detailed material (a procedure with many parts and steps, such as assembling and starting up a new com-

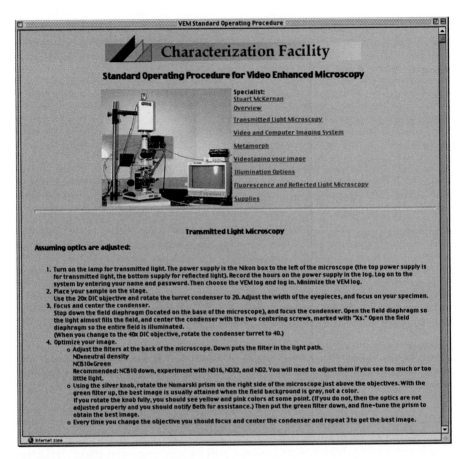

Figure 11.6 ■ A standard operating procedure. This document is part of a set of instrument instructions for the University of Minnesota's Center for Interfacial Engineering. This SOP is available on a Web site, which makes it readily accessible to readers at different physical locations.

Source: Reprinted with the permission of the Regents of the University of Minnesota.

puter), you would want to create long instructions. These instructions might be turned into a manual (see the next section), a Web site, or a CD-ROM. Figure 11.7 is a portion of a set of long instructions on installing glass block. Note the use of bullets to list the parts users will need, the list of steps, and the clear illustrations.

Typical components of procedures

Title. Create a brief, succinct title that clearly states what this procedure is about; for example, "Setting Up Your Power Macintosh G3."

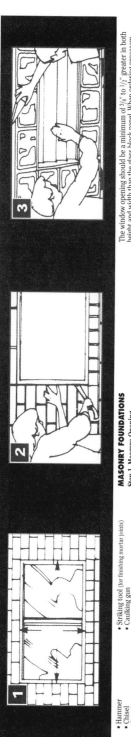

- Hammer
- Chisel
- Crowbar
- Trowel
- Premixed glass block mortar
- Hacksaw or wood saw
- Striking tool (for finishing mortar joints)
- Caulking gun
- Clear or colored caulk
- Wood or aluminum molding(s)
- Expansion strip material
- Gloves

For positioning panels in masonry openings, you'll need wood wedges — 3-1/2" long, tapered from 1-1/2" to 0" in height. To hold the panel in place in wood frames, you'll need wood shims. And, for reducing oversize openings in wood frames, you'll need lengths of 1" x 4" or 2" x 4" lumber (See Fig. A).

- Martillo
- Cincel
- Palanca de pie de cabra
- Paluste
- Guantes
- Nivel
- Sierra para metal o para madera
- Calalateo transparente o a color
- Pistola para calatateo
- Molduras de madera o aluminio
- Herramienta de formación (para acabado del mortero)
- Material para tiras de expansión
- Mortero premezclado parabio que de vidrio

Para colocar los paneles en los orificios de la mampostería, necesitará cuñas de madera de 3,5 pulgadas (8,9 cm) de largo, ahusadas de 1,5 a 0 pulgadas (4 a 0 cm) de alto. Para sujetar el panel en su sitio dentro de los marcos de madera, necesi-

- marteau
- pince à décoffrer
- barre à mine
- truelle
- ciseau à bois
- niveau
- mortier pré-mélangé spécial brique de verre
- scie à métaux ou à bois
- pistolet à mastic
- mastic transparent ou de couleur
- outil de bourrage des joints
- baguette(s) en bois ou en aluminium
- bande de dilatation
- gants

Pour bien positionner un panneau dans une ouverture en maçonnerie, servez-vous de coins en bois biseautés d'environ 3 1/2 pouces (90 mm) de long et d'é-paisseur maximale de 1 1/2 pouces (40 mm). Si le bâti est en bois, servez vous de

MASONRY FOUNDATIONS

Step 1. Measure Opening

Before ordering a glass block panel, carefully measure the opening from the *outside* of the house.

Measure the *full* width of the masonry opening, *as if the existing frame were already removed*. Measure the height of the opening from the top of the foundation wall (at the sill bottom) up to the wood or steel supporting member. This height dimension should *include* the sill thickness. Most sills, especially sloped ones, are removable. To be sure, tap it — if it sounds hollow, it should come off easily when you're ready to install the new window. If the sill appears *permanent*, take the height dimension from sill top (See Fig. 1).

tará pequeñas cuñas de madera. Y, para reducir las aperturas demasiado grandes en los marcos de madera, necesitará trozos de madera de 1 x 4 pulgadas (de 2,5 a 10,2 cm) de 2 x 4 pulgadas (de 5 a 10,2 cm) (Vea la Fig. A).

CIMIENTOS DE MAMPOSTERIA

Etapa 1. Mida el orificio

Antes de pedir un panel de bloque de vidrio, mida bien el orificio desde el exterior de la casa.

Mida todo el ancho del orificio en la mampostería, *como si ya se hubiera removido el marco existente*. Mida la altura desde la parte superior del muro de cimientos (desde la base del batiente) hasta la pieza de soporte de madera o acero. Esta medición de la altura debe *incluir* el grosor del batiente. La mayoría de los

cales minces en bois afin de maintenir le panneau en bois. Si vous devez adapter un panneau de verre à une ouverture en bois surdimensionnée, utilisez des tasseaux de 1 x 4 ou de 2 x 4 pouces (25 x 100 ou 50 x 100 mm) (Figure A).

BÂTI EN MAÇONNERIE

Étape 1: mesure de l'ouverture

Avant de commander un panneau de briques de verre, mesurez avec soin les dimensions de l'ouverture devant le recevoir en vous plaçant à l'*extérieur* de la maison. Mesurez *nom largeur totale* de l'ouverture, *comme si la fenêtre avait déjà été retirée*. Mesurez ensuite sa hauteur entre la base de l'appui et le linteau porteur en bois ou en acier. La hauteur mesurée *devra inclure* l'épaisseur du seuil de la fenêtre en place. Dans la plupart des cas, ce seuil, notamment lorsqu'il est

The window opening should be a minimum of 3/8" to 1/2" greater in both height and width than the glass block panel. When ordering preassembled panels, be sure to specify "opening size." Your local retailer or fabricator will recommend the size which best fits your opening.

Step 2. Remove Old Window

You've made your preparations and the glass block panel is at hand. Now you're ready to start the actual work by removing the old window.

First, working from the *inside*, take the *movable* part of the window out of the frame.

Then, from the *outside*, pry the bottom of the frame away from the sill by using a crowbar at the middle of the frame bottom (place a piece of

batientes, especialmente los inclinados, son removibles. A fin de estar seguro, dele un golpecito / si suena hueco, debe salir fácilmente cuando esté listo para instalar la ventana. Si el batiente parece ser *permanente*, tome la medición desde la parte superior del batiente (Vea la Fig. 1).

El orificio para la ventana debe ser como mínimo de 3/8 a 1/2 pulgada (de 10 a 13 mm) más alto y ancho que el panel de bloque de vidrio. Al pedir paneles pre-fabricados, no olvide especificar el tamaño del orificio." El detallista de la localidad o el fabricante le recomendará el tamaño más apto para el orificio.

Etapa 2. Remueva la ventana vieja

Ya se ha preparado y tiene a mano el panel de bloque de vidrio. Ahora está listo para remover la ventana vieja.

biseauté, peut être supprimé. Testez-le en frappant légèrement dessus : s'il sonne creux, il devrait être facile à retirer au moment de l'installation du panneau. Si le seuil de la fenêtre paraît fixe, mesurez alors la hauteur de l'ouverture sans l'incure (Figure 1).

Les dimensions de l'ouverture devraient être d'au moins 3/8 à 1/2 pouces (10 à 13 mm) supérieures en hauteur et en largeur à celles du panneau en briques de verre. Lorsque vous passerez la commande de votre panneau préfabriqué, pré-cisez-bien que vos dimensions sont celles de l'ouverture telle que vous venez de la mesurer. Votre fournisseur ou fabricant vous recommandera alors la taille du panneau qui correspond le mieux à vos dimensions.

Étape 2: retrait de l'ancienne fenêtre

FiGURE 11.7 ■ Long instructions for installing glass block. Note that each page contains information in English, Spanish, and French. The use of color helps separate each language section.

Source: Pittsburgh Corning Corporation.

Overview. For longer procedures or those designed for audiences who are new to the task, provide a brief overview of how to use the document and what the document contains.

List of steps. With few exceptions, procedures require audiences to follow explicit steps. Make your steps clear, and number them with large enough type so readers can easily return to a particular step if they look away from the document.

Warnings, cautions. Procedures routinely include advice about avoiding injury or equipment damage. In technical writing, safety information is generally conveyed through four types of information. These are listed in order from least to most serious:

Note. Clarifies a point, emphasizes vital information, or describes options or alternatives.

> NOTE: If you don't name a newly initialized disk, the computer automatically names it "Untitled."

Caution. Prevents possible mistakes that could result in injury or equipment damage.

> CAUTION: A momentary electrical surge could erase the contents of your working document, so make sure you back up your data.

Warning. Alerts users to potential hazards to life or limb.

> WARNING: To prevent electrical shock, always disconnect your printer from its power source before you clean any internal parts.

Danger. Identifies an immediate hazard to life or limb.

> DANGER: The red canister contains DEADLY radioactive material. Do not break the safety seal under any circumstances.

You can visually emphasize these items with hazard symbols as shown below.

| Warning | Do not enter | Radioactivity | Fire Danger |

Procedure number and revision dates. Many procedures, especially SOPs, are given procedure numbers so that users can refer to a particular document. Instead of asking for the SOP about waste disposal (which could mean one of many documents), an engineer can ask for SOP #35.2. The number before the period (35) indicates the SOP itself, and the number after the period (2) indicates the revision. Many SOPs also list the revision date somewhere on the document itself.

Usability considerations

Understand the physical location where people will use the information. Will people use the procedures out on the factory floor? If so, your materials may need to be in large type to allow for less than perfect lighting. You may also need to think about using plastic-coated pages in a binder or other ways to protect the material in certain settings. If users of the document are located in an office, working at a computer, for example, you may want to design a document that stays open easily (by using comb binding, for example) so people can read and type at the same time. In other cases, you may want to put the instructions directly on the equipment.

Understand the purpose and tasks for this document. Go back to your audience and purpose analysis and be sure you are clear on how and why your audience will use this document. If workers need to refer to the standard operating procedures while in the middle of a task, they won't want to search for the information. Therefore, headings should be clear, steps should be listed in order, and each page should be visually accessible.

Understand the technical expertise of your audience. How much background do your users have? Experienced users need only new information, while novice users need help getting started. In addition, individual readers may scrutinize the document more carefully, while teams may pass the document back and forth, jumping from item to item. Especially for teams, make sure your document is easy to read and steps are clearly numbered.

Test your documents. Make sure you test the usability of your procedures on a small group of users. Procedures often address serious safety issues, so test your document to be sure people are using it as you intended (see Chapter 3 for more information on usability testing).

Use active voice. For the part of the procedures that gives specific instructions, use active voice and imperative mood. For example, write "Insert the bolt," "Remove the wiring insulation," or "Insert the floppy disk."

Organize content chronologically. List the steps in the order you want users to perform them. Number the steps, and refer back to individual steps by number, not by content ("Return to Step 4 and repeat this function.")

Documentation and manuals

The situation

Documentation and manuals are used in many of the situations and document types described here and in Chapter 10. Documentation may include brief instructions, procedures, and in descriptions. It may also describe processes and provide background on the product. Some documentation comes as a set of manuals. In a large sense, documentation is meant to do what its name implies: to *document* (provide all the supporting information for) a scientific or technical product, suite of products (such as a multimedia system consisting of many pieces of hardware and software) or services.

Documentation is called for when you have a product that is complex and requires users to have a broad range of information. Often, documentation is written in a *layered* format (Chapter 3): a large manual with all the technical details for programmers and high-end users, a quick reference card for those who just need reminders of the main keys and tasks, and a "getting started" brochure for people who want to jump right in.

More and more, documentation is produced online. If you buy a new computer, it often comes with a thin instruction manual, which contains just enough information to get you started. Once you have the machine up and running, you can access the entire library of information through the system's online help screens. Electronic documentation and manuals can be superior to paper because users can search for terms quickly without paging through an index or table of contents. In addition, online information is often *context-sensitive*: If you need help in the middle of a task (trying to save a file, for example), you can often get help with that task as you are attempting to perform it (Chapter 5).

Although the term "documentation" refers to many types of documents, this section focuses on manuals.

Audience and purpose analysis

Determine the audience's technical background and level of familiarity with the product. Often, a company may choose to produce a manual as part of a library of information products to accompany the main product. For example, if your company writes business software for networked computers, your audience is vast. You may discover that your audience includes the following groups:

- *Network administrators,* who need to know the product's technical specifications
- *Managers,* who need to understand the bottom line, business details of the software
- *Sales representatives,* who need a more detailed understanding of the product so they can use it to generate quotes and keep track of customers

Your first task would be to determine which of these users requires which type of documentation. Network administrators may need the largest manual, because they will need to troubleshoot an entire range of problems, from user errors to system crashes.

In terms of purpose, learn all you can about how your audience will use the document and what tasks they need to perform. After conducting interviews and studying the daily activities of network administrators, for example, you may discover that they spend 50% of their time looking up user errors, 10% checking out potential system errors, and 40% updating the system and running reports. This analysis should guide you in creating a manual that focuses on user errors, followed in importance by system issues.

Types of manuals

A complete manual contains background information, specifications, descriptions, and procedures, all in one document. This type of manual is appropriate for a situation or product that is not very complex—a home appliance, computer peripheral (scanner or printer), or a simple software application. For more complex products or services, a single document that contained all information would be too large and would be hard for mixed audiences to access: Novices would find leafing through the specifications part frustrating, and experts would find all the getting started information superfluous.

Increasingly common, especially for computer equipment, is a more concise type of manual. Often printed in small book format, the concise manual might offer computer startup instructions, brief specifications, and basic operating tips. The more lengthy and helpful information is contained within the computer software (online help), and users can access this information once they've started up the machine.

Typical components of manuals

Overview. Since manuals are often large documents, it is important to provide your audience with a road map before they get started. An overview section can be brief, and it should outline what the manual contains and how users should approach the document.

Access points. Users may only need to locate one piece of information in a large manual, so make sure your access points—table of contents (Figure 11.8), running heads, index—are well developed and easy to use. Test them to see if users can actually find what they are looking for.

Chapters. In longer manuals, divide the material into chapters. Each chapter contains a logical grouping of information: "Getting Started," "Installing the Printer," "Connecting to the Network," and so on. Order chapters in the sequence you want users to encounter the information. If the manual contains

---

CONFLICT CATCHER TABLE OF CONTENTS   v

CONTENTS

CHAPTER 1: WELCOME TO CONFLICT CATCHER... 3

What Conflict Catcher Does 3

Extensions: a Crash Course 4

Installing Conflict Catcher 5

What's New in Conflict Catcher 8 7

CHAPTER 2: MANAGING YOUR EXTENSIONS AND CONTROL PANELS.....11

How to Open Conflict Catcher 11

The Conflict Catcher Window 12

Everyday Conflict Catcher 22

The Conflict Catcher Columns 23

Fun Things to Do During Startup 28

Reports 30

Online Help 32

CHAPTER 3: EXTENSIONS EN MASSE: SETS AND LINKS................................... 37

Sets: Pre-defined File Lists 37

Links: Clusters of Related Files 49

CHAPTER 4: CONFLICT CATCHING 57

How Conflict Catcher Catches Conflicts 58

Conducting a Startup Test 59

Crashes At Startup: Automatic Conflict Testing 71

What to Do about Problem Files 73

CHAPTER 5: MANAGING SPECIALTY FOLDERS 77

Managing Fonts, Plug-ins, and Other Files 77

FIGURE 11.8 ■ **Table of contents from the *Conflict Catcher* manual.**

Source: Courtesy of Casady and Greene.

more than roughly 10 chapters, you might group the chapters into sections: "Section One: Getting Started," "Section Two: For Network Administrators," "Section Three: Specifications." Each section would then contain approximately 3 to 6 chapters.

Reference information. No manual can contain everything every user will need. So make sure you provide a Web address, email address, and phone number for users to contact someone if they get stuck or need more information.

Usability considerations

Determine the appropriate medium. For large-scale computer systems, documentation may be delivered in many ways. For example, it may be printed and bound into several volumes, often in three-ring binders. Each binder may contain a specific category of information: user error codes, system messages, and so on. In addition, this information may also be delivered via the Web or on a CD. If you can, use the document type that is most familiar to the audience. And if you need to upgrade (for example, if your company determines that the printed format becomes outdated too quickly and the information should be delivered entirely via the Web), make sure you provide plenty of transition time for the users. Allow users to have their paper manuals and the new electronic information at the same time until they become familiar with the new medium.

Also, if you intend to use both paper and electronic documentation, plan them at the same time to make sure the terminology and information are consistent.

Understand what information your audience does and doesn't need. In the old days of writing manuals, writers tended to include every detail about the product. Often, manual writers were engineers who worked closely on the project, and so were understandably eager to explain all the technical details, the history of the project, and so on. While this might be interesting to some readers, it is not the sort of information end users generally need.

Understand the physical location where people will use the information. If users of the document are working at a computer in an office, for example, you may want to design a manual that stays open easily so people can read and type at the same time. If your audience is network administrators, who are usually able to access the system when they encounter a problem, consider placing the information online.

Write from the user's point of view. Manuals are for users, and they should be written to reflect user needs. Instead of headings that are abstract, such as "Installation of the Microprocessor," try writing from the user's point of view. Headings in the form of questions that a user might actually ask, such as "How do I install the microprocessor?" can be very helpful. Also, make sure your material reflects what users really *care* about. If a long history of the product is not appropriate for this audience, omit it.

Include diagrams, screen samples, and illustrations. Since manuals can be long documents, it's helpful to break up the text with visuals (Figure 11.9). Visuals can also provide users with a clearer idea of what to do or what to look for on the screen.

Adopt a style that is appropriate for your audience. Most times, a neutral, informative style is most appropriate for a manual. However, for certain subjects, a more informal, energized style may be useful. Consider the following examples from the *Conflict Catcher* user manual. (*Conflict Catcher* software helps Macintosh users solve problems with system crashes, freeze-ups, and other technical difficulties.)

> "Conflict Catcher's ability to *tame these little programs* can have considerable impact on your daily Macintosh life." (p. 3)

> "On the other hand, there's *no need for paranoia*. When a System Folder is truly corrupted, the problematic file is frequently the System Suitcase file." (p. 96)

Note the upbeat, almost comic, ring to these sentences. "Tame these little programs" and "no need for paranoia" are not standard phrases in technical writing. But in the case of this software, the writer probably knew that users who need *Conflict Catcher* are already in a bad mood because of computer trouble. The casual tone is designed to calm users down and get them to focus on the material so they can solve the problem.

TECHNICAL MARKETING MATERIAL

The situation

Technical marketing material is designed to persuade an audience to purchase a product or service. Unlike proposals (Chapter 12), which are also used to sell a product or service, technical marketing materials tend to be less formal and more dynamic, colorful, and varied. A typical proposal is tailored to one client's specific needs and follows a fairly standard format; marketing literature, on the other hand, seeks to present the product in its best light for a broad array of audiences and needs. And, unlike nontechnical sales material, technical marketing documents are designed for science and technology products and are often aimed at knowledgeable readers. A team of scientists looking to

Solving your computer problems with Norton Utilities

2   Select the drive you want to diagnose.

Tip: It is okay to select more than one drive to diagnose.

3   Click Diagnose.

Norton Disk Doctor checks the various components of your disk.

4   Follow the on-screen prompts as Norton Disk Doctor identifies and fixes any problems found on your disk.

Fixing Windows problems

To find and fix Windows problems using Norton WinDoctor:

1   Click the Start button, and then select Programs > Norton System Works > Norton Utilities > Norton WinDoctor.

2   Follow the Norton WinDoctor Wizard instructions to check your system for problems.

3   Click Finish.

Norton WinDoctor displays a checklist of the problems it found.

4   You can choose to fix all or some of the problems.

- Click Repair All to automatically correct all problems.

- Select a specific problem and click Repair.

Sub-problems for the high-level problems selected in the Problems Found list

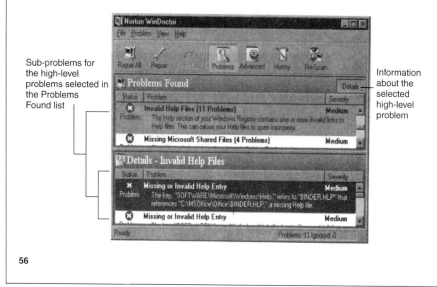

Information about the selected high-level problem

56

FiɢuRᴇ 11.9 ■ **Screen shots used along with text.**

Source: Reprinted with permission from Symantec Corporation.

purchase a new electron microscope wants marketing material that provides specific technical information. Even for a general audience (say, home computer users), technical marketing material must deal with specialized concepts not ordinarily found in nontechnical sales materials.

Technical communicators with a flare for the creative are often hired as technical marketing specialists. Some situations that call for technical marketing material include the following:

- *Cold calls*—sales representatives sending material to a range of potential new customers
- *On-site visits*—sales representatives and technical experts visiting a customer to see if a new product or service might be of interest
- *Display booths*—booths at industry trade shows displaying engaging, interesting material that people can take and read at their leisure
- *Web information*—Web pages acting as the primary place users go to find information on a technical product or service

Audience and purpose analysis

When creating marketing materials, you need a clear sense of your audience. Who are the readers of these materials: managers? technical experts? purchasing agents? The level of technical language you use will be based on the answers. Also, make sure you know how audiences will use the documents. The main purpose of most marketing materials is to make potential customers aware of your product or service and to sell them your particular company, brand, or model. Marketing materials are essentially persuasive documents.

Types of technical marketing material

Brochures. The term "brochure" covers many types of documents. A typical brochure is a standard size piece of paper (8-1/2 x 11) folded in thirds, but a brochure can be designed in many sizes, depending on purpose, audience, and budget. Brochures are used to introduce a product or service, provide pricing information, and explain how customers can contact the company.

Web pages. Many companies are using the Web for their marketing materials. The advantage of a Web page (Figure 11.10) over a printed document is that if you change the price, specifications, version, look and feel, or other features of the product, you can easily make immediate changes to the Web site. Web pages also allow you to build in interactivity: Customers can give you feedback, request additional information, or place orders.

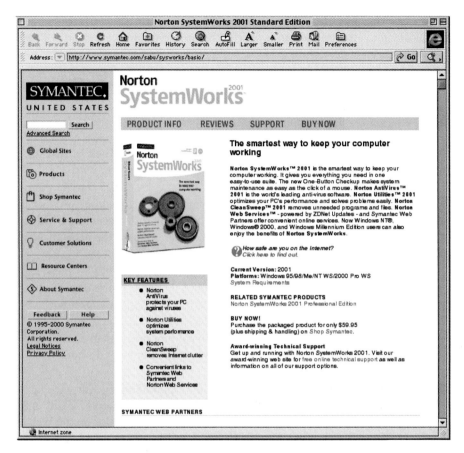

Figure 11.10 ■ **Web-based marketing material.** Many companies, especially those that produce software or computer products, make their marketing material available on the Web. Web information can be easily updated when features, versions, or prices change. In addition, this Web page follows the color scheme (yellow and black) of the product's packaging.
Source: Reprinted with permission from Symantec Corporation.

Letters. Letters can also be used for technical marketing. If a potential customer requests details about a product or service, you may send this information along with a brief cover letter. The letter should not only thank the customer for his or her interest, but also point out the specific features of your product or service that match this customer's needs.

Large color documents. Some companies produce technical marketing material that is far more comprehensive than a typical brochure. If you've ever looked at the glossy booklets for a new car, you've seen how these large color documents serve to market a product. The high-quality photography, slick color printing, and glossy feel are designed to evoke the feeling of owning a new car. These booklets also include technical specifications, such as engine horsepower and wheel base size.

Typical components of technical marketing material

Name of product. Most technical marketing material is designed to explain a product, so the name of this product should be clear and prominent.

Category or type of product or service. It's important to situate your product in relation to others of its class. For example, a brochure for a pacemaker will explain that this product is a medical device and will compare the pacemaker to other similar medical devices.

Features. Technical marketing material should describe the product's main features and explain the distinguishing features of this specific product.

Technical specifications. Many types of technical marketing materials provide specifications: product size, weight, electrical requirements, and so on.

Visuals. Visuals, especially diagrams and color photographs, are extremely effective in marketing materials. On the Web, you can use color easily and efficiently: Four-color brochures are expensive to print, but color on the Web is simply a matter of using the correct HTML tag.

Frequently Asked Questions (FAQs). Some marketing materials attempt to answer customer questions with a "frequently asked questions" (FAQ) section.

Usability considerations

Learn as much as you can about the background and experience of decision makers. Although your materials may be read by a range of people, your main goal is to get the attention of those who make the final policy and purchasing decisions. So your brochure, Web site, or other material should be geared toward their level of expertise and needs. For example, the medical device brochure in Figure 11.11 uses technical language ("steroid eluting") for its audience of medical professionals, while the car battery brochure in Figure 11.12 uses simpler language for a more general audience. Also, if you know that the decision makers value a product's effectiveness over its cost, emphasize quality, not price. For example, although cost is always a factor, physicians ultimately want medical devices to work properly.

SMALL-BODIED, STEROID-ELUTING,
TINED LEAD

BIPOLAR MODELS 4092/4592 AND 5092/5592

- Proven steroid performance in more than one million leads implanted

- New silicone allows smaller size

- Medtronic offers the broadest choice of leads to meet unique patient needs and physician preferences

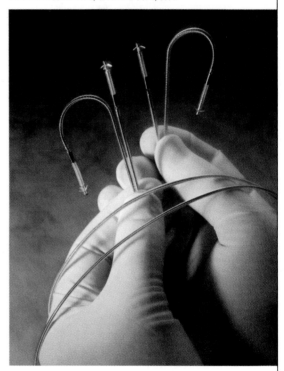

THE WORLD'S BEST SELLING LEAD
JUST GOT EVEN BETTER

Medtronic

FIGURE 11.11 ■ **A brochure for medical devices.** This cover photograph clearly shows the product and its size in relation to a human hand. The bullet points address the audience of medical professionals.

Source: Reprinted with permission from Medtronic, Inc.

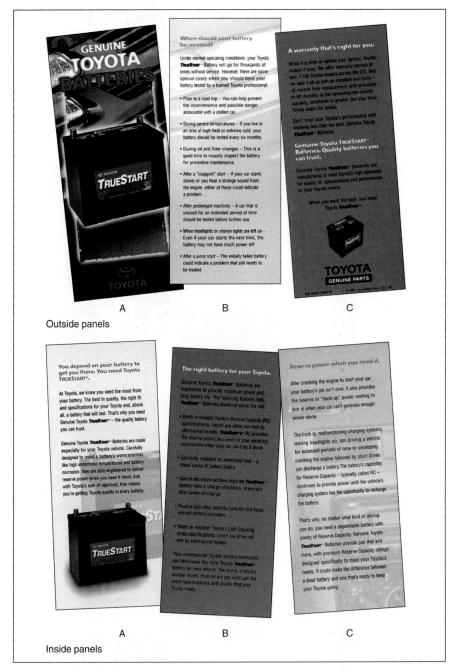

Figure 11.12 ■ A three-panel, foldout brochure for a general audience. The three-dimensional illustration in outside Panel A enables readers to visualize the product immediately. Inside Panel B—at the brochure's very center and highlighted by a blue screen—convincingly supports Panel A's claim by listing the features and specifications that make this battery special.

Source: Courtesy of Toyota Motor Sales, U.S.A., Inc.

Use upbeat, dynamic language. Be careful not to overdo it, though. Technical people tend to dislike an obvious sales pitch.

Use visuals and color. Color images can accurately convey the shape and feel of the item; also, color can add excitement and visual interest to your materials (see Figure 11.10). If you create both print and Web material, make sure your color choices are consistent, so that you convey an overall look and feel for your company and product.

Emphasize the special appeal of this product or service. Briefly explain how this product or service fits the reader's exact needs and provide solid evidence to support your claim.

 REVIEW CHECKLIST

Type of Communication	Is Called for When You Need to ...	Specific Types
Specifications	prescribe standards for performance, safety, and quality and describe features such as methods for manufacturing, building, or installing a product; materials and equipment used; and size, shape, and weight	Industry standards Government standards Functional specs Internet specs
Brief instructions	provide users with quick, easy-to-use information on how to assemble, connect, and/or use a product	Quick reference cards Assembly instructions Wordless instructions
Procedures	provide information, steps, and guidelines for completing a task in a format that is longer and more comprehensive than brief instructions	Standard operating procedures (SOPs) Long instructions
Documentation and manuals	provide all the supporting information for a scientific or technical product, suite of products, or service	Complete manuals Procedures and specifications
Technical marketing material	persuade an audience to purchase a product or service	Brochures Web pages Letters Large color documents

Exercises

1. How do specifications come into play in your workplace or even your home? Find one example of specs used in your home or workplace and complete a usability analysis. For example, if you recently purchased an item that needed to be assembled, it probably came with a list of parts. How easily were you able to find the parts listed in the specs? What could the writers have done to improve the usability of this document?

2. FOCUS ON ▨▶WRITING. Find two examples of documentation, one that you consider good and one that you think is bad. These can be from any source. With one or two other students, list the features that make the good one good and the bad one bad. Now, draft a memo to the writer of the bad documentation outlining things he or she should do to improve on the next version.

The Collaboration Window

Form small groups according to your major. Discuss several major topics in your field or from a recent job or internship, and settle collectively on a topic for a short report. Complete an audience and purpose analysis, paying special attention to the features listed in this chapter. Create a detailed outline of what the report would contain.

The Global Window

Assume the following scenario: Members of your environmental consulting firm travel in teams worldwide on short notice to manage various environmental emergencies (toxic spills, chemical fires, and the like). Because of the rapid response required for these assignments and the international array of clients being served, team members have little or no time to research the particular cultural values of each client. Members typically find themselves having to establish immediate rapport and achieve agreement as they collaborate with clients during highly stressful situations.

Too often, however, ignorance of cultural differences leads to misunderstanding and needless delays in critical situations. Clients can lose face when they feel they are being overtly criticized and when their customs or values are ignored. When people feel insulted, or offended by inappropriate behavior, communication breaks down.

To avoid such problems, your boss has asked you to prepare a set of brief, general instructions titled "How to Avoid Offending International Clients." For immediate access, the instructions should fit on a pocket-sized quick reference card.

Working alone or in groups, do the research and design the reference card.

Click on This

You can learn more about standards and SOPs from the following Web sites:

www.standards.ieee.org/catalog/index.html

www.w3.org

www.lehigh.edu/ ~ kaf3/sops/sop_index.html.

An excellent guide to SOPs produced by Professor Kenneth Friedman in the Department of Journalism and Communication, Lehigh University.

CHAPTER **12**

Complex Communication
Situations

Definitions and descriptions

Long reports

Proposals

Review checklist

Exercises

The collaboration window

The global window

Click on this

Definitions and descriptions

The situation

Definitions explain a term or concept that is specialized or unfamiliar to an audience. In some cases, a term may have more than one meaning, and a clearly written definition tells audiences exactly how the term is being used. Such precision is important in technical fields, where terms and phrases usually have specific meanings. Engineers talk about "elasticity" or "ductility"; bankers discuss "amortization" or "fiduciary relationships." For people both inside and outside the fields, these terms must be defined.

Descriptions, like definitions, help define an idea. In addition, descriptions use words and visuals to create a picture of the product or process, such as the structure of a bicycle frame or the process of nuclear fusion. These strategies (definition and description) rely on each other and provide the basis for virtually any type of technical explanation. For example, an audience of electrical engineering students may need a description of a solenoid, turbine, or some other electrical component. You may also need to use description or definition for lay audiences, such as people who want to know what a circuit breaker is or how an automobile bumper jack works in case of a flat tire.

Audience and purpose analysis

Make sure your particular language and content matches the audience's background and experience. For a group of electrical engineers, your definition or description of a solenoid can be brief and to the point, and you can use highly technical language. For engineering students, your definition will need more detail. For general audiences, your definition will require language they can understand. See Figure 12.1 for a definition and description that includes a diagram to help general readers understand the entire process by which chemical contaminants are removed from soil.

In terms of purpose, you need to understand why a particular audience needs or wants this information. If you are describing an automobile bumper jack for a general audience, you may learn that their main purpose is to understand how the jack works so they can eventually use it to change a car tire. As such, the description (or definition) should be written in clear language and should avoid highly technical terms. It should follow an outline format or other layout pattern that uses headings and white space, so readers can get in and out of the document quickly. And it should include diagrams or drawings, to help readers see the whole picture. (See Figures 12.2 and 12.3 for two sample descriptions.)

Dubbed the "lasagna" process because of its layers, this technology cleans up liquid-borne organic and inorganic contaminants in dense, claylike soils. Initial work is focused on removing chlorinated solvents.

Because clay is not very permeable, it holds ground water and other liquids well. Traditional remediation for this type of site requires that the liquid in the soil (usually ground water) be pumped out. The water brings many of the contaminants with it, then is chemically treated and replaced—a time-consuming and expensive solution.

The lasagna process, on the other hand, allows the soil to be remediated *in situ* by using low-voltage electric current to move contaminated ground water through treatment zones in the soil. Depending on the characteristics of the individual site, the process can be done in either a horizontal or vertical configuration. (See figure below.)

The first step in the lasagna process is to "fracture" the soil, creating a series of zones. In a horizontal configuration, a vertical borehole is drilled and a nozzle inserted; a highly pressur-

ized mixture of water and sand (or another water/solid mix) is injected into the ground at various depths. The result: a stack of pancake-shaped, permeable zones in the denser, contaminated soil. The top and bottom zones are filled with carbon or graphite so they can conduct electricity. The zones between them are filled with treatment chemicals or microorganisms that will remediate the contaminants.

When electricity is applied to the carbon and graphite zones, they act as electrodes, creating an electric field. Within the field, the materials in the soil migrate toward either the positive or negative electrode. Along with the migrating materials, pollutants are carried into the treatment zones, where they are neutralized or destroyed.

The vertical configuration works in much the same way, differing only in installation. Because the electrodes and treatment zones extend down from the surface, this configuration does not require the sophisticated hydraulic fracturing techniques that are used in the horizontal configuration.

Schematic Diagram of the Lasagna Process

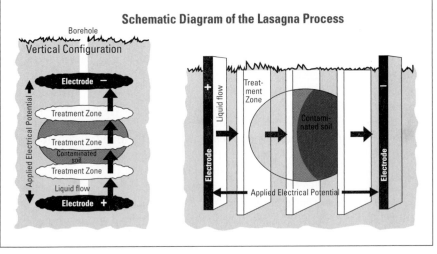

FIGURE 12.1 ■ **A definition and description of the "lasagna process" for filtering contaminants out of soil.**

Source: Adapted from Japikse, C. (1994). Lasagna in the making. *EPA Journal* 20 (3), 27.

A Description of the Standard Stethoscope

The stethoscope is a listening device that amplifies and transmits body sounds to aid in detecting physical abnormalities.

This instrument has evolved from the original wooden, funnel-shaped instrument invented by a French physician, R. T. Lennaec, in 1819. Because of his female patients' modesty, he found it necessary to develop a device, other than his ear, for auscultation (listening to body sounds).

This report explains to the beginning paramedical or nursing student the structure, assembly, and operating principle of the stethoscope.

The standard stethoscope is roughly 24 inches long and weighs about 5 ounces. The instrument consists of a sensitive sound-detecting and amplifying device whose flat surface is pressed against a bodily area. This amplifying device is attached to rubber and metal tubing that transmits the body sound to a listening device inserted in the ear.

The stethoscope's Y-shaped structure contains seven interlocking pieces: (1) diaphragm contact piece, (2) lower tubing, (3) Y-shaped metal piece, (4) upper tubing, (5) U-shaped metal strip, (6) curved metal tubing, and (7) hollow ear plugs. These parts form a continuous unit (Figure 1).

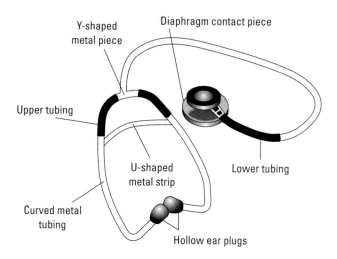

FIGURE 1 Stethoscope with Diaphragm Contact Piece (Front View)

FiɢᴜʀE 12.2 ■ A short description of a standard stethoscope.

The seven major parts of the stethoscope provide support for the instrument, flexibility of movement for the operator, and ease in use.

In an operating cycle, the diaphragm contact piece, placed against the skin, picks up sound impulses from the body's surface. These impulses cause the plastic diaphragm to vibrate. The amplified vibrations, in turn, are carried through a tube to a dividing point. From here, the amplified sound is carried through two separate but identical series of tubes to hollow ear plugs.

Figure 12.2 ■ *(Continued)*

I. Introduction: General Description

   A. Definition, Function, and Background of the Item

   B. Purpose (and Audience—for classroom only)

   C. Overall Description (with general visuals, if applicable)

   D. Principle of Operation (if applicable)

   E. List of Major Parts

II. Description and Function of Parts

   A. Part One in Your Descriptive Sequence

      1. Definition

      2. Shape, dimensions, material (with specific visuals)

      3. Subparts (if applicable)

      4. Function

      5. Relation to adjoining parts

      6. Mode of attachment (if applicable)

   B. Part Two in Your Descriptive Sequence (and so on)

III. Summary and Operating Description

   A. Summary (used only in a long, complex description)

   B. Interrelation of Parts

   C. One Complete Operating Cycle

Figure 12.3 ■ **A longer description of an automobile bumper jack.** This sample outline is also included.

Description of a Standard Bumper Jack

Introduction—General Description

The standard bumper jack is a portable mechanism for raising the front or rear of a car through force applied with a lever. This jack enables even a frail person to lift one corner of a 2-ton automobile.

The jack consists of a molded steel base supporting a free-standing, perpendicular, notched shaft (Figure 1). Attached to the shaft are a leverage mechanism, a bumper catch, and a cylinder for insertion of the jack handle. Except for the main shaft and leverage mechanism, the jack is made to be dismantled. All its parts fit neatly in the car's trunk.

The jack operates on a leverage principle, with the operator's hand traveling 18 inches and the car only $\frac{3}{8}$ of an inch during a normal jacking stroke. Such a device requires many strokes to raise the car off the ground but may prove a lifesaver to a motorist on some deserted road.

Five main parts make up the jack: base, notched shaft, leverage mechanism, bumper catch, and handle.

Description of Parts and Their Function

Base. The rectangular base is a molded steel plate that provides support and a point of insertion for the shaft (Figure 2). The base slopes upward to form a platform containing a 1-inch depression that provides a stabilizing well for the shaft. Stability is increased by a 1-inch cuff around the well. As the base rests on its flat surface, the bottom end of the shaft is inserted into its stabilizing well.

Shaft. The notched shaft is a steel bar (32 inches long) that provides a vertical track for the leverage mechanism. The notches, which hold the mechanism in position on the shaft, face the operator.

The shaft vertically supports the raised automobile, and attached to it is the leverage mechanism, which rests on individual notches.

Leverage Mechanism. The leverage mechanism provides the mechanical advantage needed for the operator to raise the car. It is made to slide up and down the notched shaft. The main body of this pressed-steel mechanism contains two units: one for transferring the leverage and one for holding the bumper catch.

The leverage unit has four major parts: the cylinder, connecting the handle and a pivot point; a lower pawl (a device that fits into the notches to allow forward and prevent backward motion), connected directly to the cylinder; an upper pawl, connected at the pivot point; and an "up-down" lever,

Figure 12.3 ■ *(Continued)*

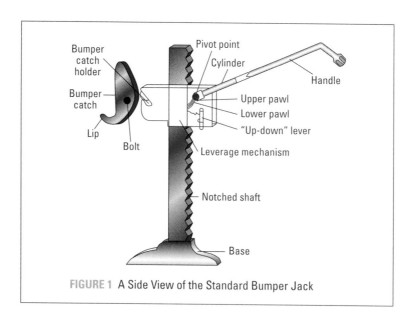

FIGURE 1 A Side View of the Standard Bumper Jack

which applies or releases pressure on the upper pawl by means of a spring (Figure 1). Moving the cylinder up and down with the handle causes the alternate release of the pawls, and thus movement up or down the shaft—depending on the setting of the "up-down" lever. The movement is transferred by the metal body of the unit to the bumper catch holder.

The holder consists of a downsloping groove, partially blocked by a wire spring (Figure 1). The spring is mounted in such a way as to keep the bumper catch in place during operation.

Bumper Catch. The bumper catch is a 9-inch molded plate that attaches the leverage mechanism to the bumper and is bent to fit the shape of the bumper. Its outer $\frac{1}{2}$-inch is bent up to form a lip (Figure 1), which hooks behind the bumper to hold the catch in place. The two sides of the plate are bent back 90 degrees to leave a 2-inch bumper contact surface, and a bolt is riveted between them. This bolt slips into the groove in the leverage mechanism and provides the attachment between the leverage unit and the car.

Jack Handle. The jack handle is a steel bar that serves both as lever and lug bolt (or lugnut) remover. This round bar is 22 inches long, $\frac{5}{8}$-inch in diameter, and is bent 135 degrees roughly 5 inches from its outer end. Its outer

FIGURE 12.3 ■ *(Continued)*

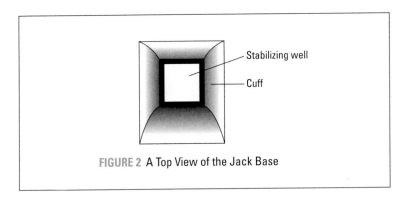

FIGURE 2 A Top View of the Jack Base

end is a socket wrench made to fit the wheel's lug bolts. Its inner end is beveled to form a bladelike point for prying the wheel covers and for insertion into the cylinder on the leverage mechanism.

Conclusion and Operating Description

One quickly assembles the jack by inserting the bottom of the notched shaft into the stabilizing well in the base, the bumper catch into the groove on the leverage mechanism, and the beveled end of the jack handle into the cylinder. The bumper catch is then attached to the bumper, with the lever set in the "up" position.

As the operator exerts an up-down pumping motion on the jack handle, the leverage mechanism gradually climbs the vertical notched shaft until the car's wheel is raised above the ground. When the lever is in the "down" position, the same pumping motion causes the leverage mechanism to descend the shaft.

Figure 12.3 ■ *(Continued)*

Types of definitions and descriptions

Brief. Often, you can clarify the meaning of a word by using a more familiar synonym or a clarifying phrase, as in

The *leaching field* (sievelike drainage area) requires crushed stone.

In an online document, such as a Web page or online help system, short definitions such as these can easily be linked to the main word or phrase. When a user clicks on "leaching field," he or she would go to a window that contains the definition and other important information.

A slightly longer way to define a phrase is to use the "term-class-features" method. You begin by stating the term. Then you indicate the broader class that this item belongs to, followed by the features that distinguish it from other items in that general grouping. For example:

Term	Class	Features
Carburetor	a mixing device	in gasoline engines that blends air and fuel into a vapor for combustion within the cylinders.
Diabetes	a metabolic disease	caused by a disorder of the pituitary gland or pancreas.

Brief definitions are fine when the audience does not require a great deal of information. For example, the sentence about the leaching field might be adequate in a progress report to a client whose house you are building. But a document that requires more detail, such as a public health report on groundwater contamination, would call for an expanded definition.

Expanded. Expanded definitions are appropriate when audiences require more detail. Depending on audience and purpose, an expanded definition may be a short paragraph or may extend to several pages. If a device, such as a digital dosimeter (used for measuring radiation exposure), is being introduced for the first time to an audience who needs to understand how this instrument works, your definition would require at least several paragraphs, if not pages. In such cases, your document would evolve from a simple definition to one that both defines and describes the device.

Typical components of definitions and descriptions

Etymology. Sometimes, a word's origin (etymology) can help users understand its meaning. "Biology," for example, is derived from the Greek "bio," meaning "life." You can use a dictionary to locate the sources of most words.

History and background. In some cases, the history or background of a term, concept, or procedure can be useful in helping to define and describe it. For students or researchers who want in-depth information, history and background is appropriate. However, for users trying to perform a task, history and background can be cumbersome and unnecessary. If you wanted to install a new modem, you might be interested in a quick sentence explaining that "modem" stands for "modulator-demodulator." But you would not really care about the history of how modems were developed.

Operating principle. If part of the document's purpose is to teach people to use a product correctly, it is usually helpful for your audience to understand how a device operates. For example, a manual for a garden rototiller is intended to help people use the tiller. Therefore, users should have a sense of how the tiller operates: it uses gasoline, it needs a clean spark plug, and so on. A list of parts is another way of illustrating how a device operates.

Usability considerations

Use appropriate levels of technicality. Your language in a definition or description needs to match the particular audience's level of experience. An audience of medical technicians will easily understand jargon related to their field, but nonexperts will need language they find familiar. For example, the sentence

❙   A tumor is a neoplasm.

would make sense to most medical professionals. But for an audience outside that field, you would need to unpack the term "neoplasm" and use more accessible language, as in

❙   A tumor is a growth of cells that occurs independently of surrounding tissue and
    serves no useful function.

Consider length and placement. The length of a definition or description should be appropriate for your audience and purpose. For example, if your audience needed to know only the very basics about a term (such as "tumor," above), you could write a short sentence. But if your audience needed more information, you would need to amplify your definition with a description (say, of the process by which tumor cells displace healthy tissue).

Placement is also important. Each time an audience encounters an unfamiliar term or concept, it should be defined or described in the same area on the page or screen. In a printed text, you can accomplish this by placing brief definitions in an outside margin. On the screen, you would use a hypertext link. Hypertext and the Web are perhaps the best answer yet to making definitions and descriptions accessible, because readers can click on the item, read about it, and return to their original place on the page.

Use visuals. Visuals can be very important in definitions and descriptions. You can explain as clearly as you like, but as the saying goes, a picture is often worth a thousand words—even more so when used with clear, accurate prose. The cutaway diagram (top view) in Figure 12.4 is a description of an electricity meter, designed for a general audience.

Use clear, concise language. Use sentences that are brief and to the point. Provide readers with the most important information quickly. If all your audience

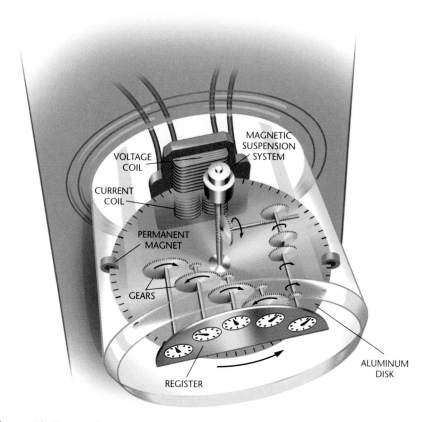

Figure 12.4 ■ **A description of an electricity meter.** This exploded diagram will help make the written description come to life.
Source: From Rosenau, L. (2000, March). Working knowledge: Electricity meters. *Scientific American,* 108.

needs is a one sentence definition, don't write a long description, no matter how interesting you think it might be.

Choose words with precision. Choose words carefully, and use the same word to refer to the same item. Is a tumor a "growth of cells" or simply "cells"? When describing fiber optic technology, don't suddenly switch and call it "high speed cable."

Use comparisons and examples. By comparing new information to ideas your audience already understands, you help build a bridge between their current knowledge and the new ideas. For example, for a group of nonexperts, you could explain how earthquakes start in this manner:

> Imagine an enormous block of gelatin with a vertical knife slit through the middle of its lower half. Giant hands are slowly pushing the right side forward and the left side back along the slit, creating a strain that eventually splits [the block].

Use an appropriate organizational sequence. For longer descriptions, choose the organizational pattern that is most consistent with your purpose. If you want to describe how something looks or what parts it has, use a *spatial* sequence: Describe the items as your audience will see them. If you want to describe how something works, use a *functional* sequence: Describe the workings (functions) of the device. And if you want to describe how something is assembled, use a *chronological* sequence (see also Chapter 11, "Brief Instructions"; "Procedures").

Long Reports

The situation

When your purpose is to inform an audience, offer a solution to a problem, report progress, or make a detailed recommendation, you may need to write a long report. Long reports are often structured like a small book, with a table of contents, appendices, and an index. As with short reports (Chapter 10), long reports present ideas and facts to interested parties, decision makers, and other audiences. Technical professionals rely on reports as a basis for making informed decisions on a range of matters, from the possible side effects of a new pain medication to the environmental risks posed by a certain gasoline additive.

Long reports are called for in situations where an audience needs detailed information, statistics, and background information—the whole story. For example, your team of engineers needs to make far-reaching decisions about the best site for a toxic waste containment field. You have several months to research and make a decision, so you hire a consulting firm to report on all the relevant information. Their resulting product, a long report describing the geologic conditions of potential sites, might contain an appendix with detailed comparisons of topsoil, ground water, and other conditions.

Audience and purpose analysis

Do your best to determine who will read the report. For instance, even if the report is addressed to team members, it may be sent on to other managers, the legal department, or sales and marketing. If you can learn about the actual audience members in advance, you can anticipate their various needs as you create the report. Before you start the report, be clear about its true purpose. For example, you may be under the impression that the report is intended simply to inform an audience. But after some initial research, you learn that your manager really wants you to recommend an action, not just state the facts. Recommending is different from informing, so it's important to understand the reason you are writing the report in the first place.

Types of long reports

Causal. Causal reports are used in situations where you need to explain what caused something to happen. For example, medical researchers may need to explain why so many apparently healthy people have sudden heart attacks. Or you might need to anticipate the possible effects of a particular decision, say, the effects of a corporate merger on employee morale.

Comparative. Comparative reports are used when you need to rate similar items on the basis of specific criteria. For example, you may need to answer questions such as "Which type of security procedure (firewall or encryption) should we install in our company's computer system?"

Feasibility. Feasibility reports are used when your purpose is to assess the practicality of an idea or plan. For example, if your company needs to know whether increased business will justify the cost of an interactive Web site, you would need to do some research and describe the results in a feasibility report.

Sometimes, these categories overlap. Any single study may in fact require you to take several approaches. The sample report in Figure 12.5 is an example. It is designed to answer two questions: "Is technical marketing the right career for me?" (feasibility), and "If so, which is my best option for entering the field?" (comparative).

Feasibility Analysis of a Career in Technical Marketing

INTRODUCTION

The escalating cutbacks in aerospace, defense-related, and other goods-producing industries have narrowed career opportunities for many of today's science and engineering graduates. A study by the Massachusetts Institute of Technology, for example, found that leading industries hired 80 to 90 percent fewer engineers in the mid-1990s than in the mid-1980s (Solomon, 1996). This trend is expected to continue—if not worsen—in the foreseeable future.

With the notable exception of computer engineering, employment in all engineering specialties will grow at rates ranging from average to far below average to near static through the year 2006. In some specialties (e.g., mining and petroleum engineering), employment will actually decline ("Occupational Employment," 1998, p. 16).

Given such bleak employment prospects, recent graduates might consider alternative careers in which they could apply their technical training. One especially attractive alternative is in marketing and selling of technology products or services.

Customer orientation is an ever-growing part of today's business and manufacturing climate. Beginning in the 1980s, U.S. industry ceased to be "manufacturing driven" (where customers would buy whatever products were available). Technology companies became "service driven" by the demand for customized, efficiently serviced products (Basta, 1984, p. 84).

In the product-oriented industries of the 1970s, technical marketing accounted for only 39 percent of top management backgrounds, but that number nearly doubled by 1995. Also, 1998 employment listings for recent graduates showed countless major companies offering positions in technical marketing and sales (College Placement Council, 1998, p. 402).

Undergraduates interested in this career need answers to these basic questions:

- *Is this the right career for me?*
- *If so, how do I enter the field?*

To help answer these questions, this report analyzes information gathered from professionals as well as from the literature.

After defining *technical marketing,* the following analysis examines the field's employment outlook, required skills and personal qualities, career benefits and drawbacks, and various entry options.

Figure 12.5 ■ **An analytical report.** This long report combines two approaches: feasibility and comparisons.

COLLECTED DATA

Key Factors in a Technical Marketing Career

Anyone considering technical marketing needs to assess whether this career fits his or her interests, abilities, and aspirations.

THE TECHNICAL MARKETING PROCESS. Although the terms *marketing* and *sales* are often used interchangeably, technical marketing traditionally has involved far more than sales work. The process itself (identifying, reaching, and selling to customers) entails six major activities (Cornelius & Lewis, 1983, p. 44):

1. *Market research:* gathering information about the size and character of the target market for a product or service.
2. *Product development and management:* producing the goods to fill a specific market need.
3. *Cost determination and pricing:* measuring every expense in the production, distribution, advertising, and sales of the product, to determine its price.
4. *Advertising and promotion:* developing and implementing all strategies for reaching customers.
5. *Product distribution:* coordinating all elements of a technical product or service, from its conception through its final delivery to the customer.
6. *Sales and technical support:* creating and maintaining customer accounts, and servicing and upgrading products.

Fully engaged in all these activities, the technical marketing professional gains a detailed understanding of the industry, the product, and the customer's needs (Figure 1).

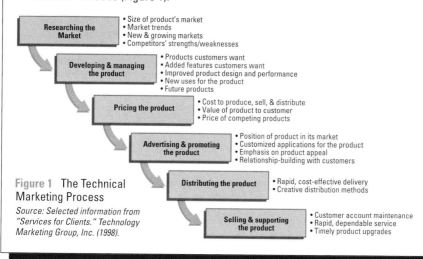

Figure 1 The Technical Marketing Process

Source: Selected information from "Services for Clients." Technology Marketing Group, Inc. (1998).

Figure 12.5 ■ *(Continued)*

EMPLOYMENT OUTLOOK. For graduates with the right combination of technical and personal qualifications, the employment outlook for technical marketing appears excellent. While engineering jobs will increase at barely one half the average growth rate for jobs requiring a Bachelor's degree, marketing jobs will exceed the average growth rate (Figure 2).

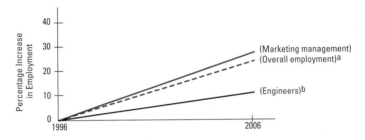

Figure 2 The Employment Outlook for Technical Marketing

[a]Jobs requiring a Bachelor's degree.
[b]Excluding outlying rates for engineering specialties at extreme ends of the spectrum (computer engineers: +109%; mining and petroleum engineers: –14%).
Source: Data from U.S. Department of Labor. (1998, Winter). Occupational Outlook Quarterly, *11–16.*

Although highly competitive, these marketing positions call for the very kinds of technical, analytical, and problem-solving skills that engineers can offer (Solomon, 1996)—especially in an automated environment.

TECHNICAL SKILLS REQUIRED. Computer networks, interactive media, and multimedia will increasingly influence the way products are advertised and sold. Also, marketing representatives increasingly work from a "virtual" office. Using laptop computers, fax networks, and personal digital assistants, representatives in the field have real-time access to electronic catalogs of product lines, multimedia presentations, pricing for customized products, inventory data, product distribution, and customized sales contacts (Tolland, 1999).

With their rich background in computer, technical, and problem-solving skills, engineering graduates are ideally suited for (a) working in automated environments, and (b) implementing and troubleshooting these complex and often sensitive electronic systems.

OTHER SKILLS AND QUALITIES REQUIRED. In marketing and sales, not even the most sophisticated information can substitute for the "human factor": the ability to connect with customers on a person-to-person level (Young, 1995, p. 95).

FɪɢᴜRE **12.5** ■ *(Continued)*

One senior sales engineer praises the efficiency of her automated sales system, but thinks that automation will "get in the way" of direct customer contact. Other technical marketing professionals express similar views about the continued importance of human interaction (94).

Besides a strong technical background, marketing requires a generous blend of those traits summarized in Figure 3.

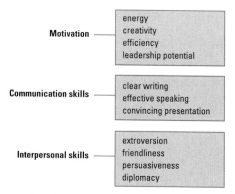

Figure 3   Requirements for a Technical Marketing Career

Motivation is essential for marketing work. Professionals must be energetic and able to function with minimal supervision. Career counselor Phil Hawkins describes the ideal candidates as people who can plan and program their own tasks, who can manage their time, and who have no fear of hard work (personal interview, February 11, 1999). Leadership potential, as demonstrated by extracurricular activities, is an asset.

Motivation alone provides no guarantee of success. Marketing professionals are paid to communicate the virtues of their products or services. This career therefore requires skill in communication, both written and oral. Documents for readers outside the organization include advertising copy, product descriptions, sales proposals, sales letters, and user manuals and online help. In-house writing includes recommendation reports, feasibility studies, progress reports, memos, and email correspondence (U.S. Department of Labor, 1997, p. 8).

Skilled oral presentation is vital to any sales effort, as Phil Hawkins points out. Technical marketing professionals need to speak confidently and persuasively—to represent their products and services in the best possible light (personal interview, February 11, 1999). Sales presentations often involve public speaking at conventions, trade shows, and other similar forums.

FIGURE 12.5 ■ *(Continued)*

Beyond motivation and communication skills, interpersonal skills are the ultimate requirement for success in marketing (Solomon, 1996). Consumers are more likely to buy a product or service when they like the person selling it. Marketing professionals are extroverted, friendly, and diplomatic; they can motivate people without alienating them.

ADVANTAGES OF THE CAREER. As shown in Figure 2, technical marketing offers diverse experience in every phase of a company's operation, from a product's design to its sales and service. Such broad exposure provides excellent preparation for countless upper-management positions.

In fact, sales engineers with solid experience often open their own businesses as "manufacturers' agents" representing a variety of companies. These agents represent products for companies who have no marketing staff of their own. In effect their own bosses, manufacturers' agents are free to choose, from among many offers, the products they wish to represent (Tolland, 1999).

Another career benefit is the attractive salary. Marketing professionals typically receive a base pay plus commissions. According to John Turnbow, managing recruiter of National Electric's technical marketing program, new marketing engineers for NE average over $60,000 in first-year wages. Moreover, salaries often reach six figures—sometimes higher than executive salaries (personal communication, April 5, 1999).

Technical marketing is especially attractive for its geographic and job mobility. Companies nationwide seek recent graduates, but especially in the Southeast and on the east and west coasts ("Electronic sales," 1996, pp. 1134–37). In addition, the types of interpersonal and communication skills that marketing professionals develop are highly portable. This is especially important in our current, rapidly shifting economy, in which job security is disappearing in the face of more and more temporary positions (Jones, 1997, p. 51).

DRAWBACKS OF THE CAREER. Technical marketing is by no means a career for every engineer. Sales engineer Roger Cayer cautions that personnel might spend most of their time traveling to meet potential customers. Success requires hard work over long hours, evenings, and occasional weekends. Above all, the job is stressful because of constant pressure to meet sales quotas (phone interview, February 8, 1999). Anyone considering this career should be able to work and thrive in a highly competitive environment.

A Comparison of Entry Options
Engineers and other technical graduates enter technical marketing through one of four options. Some join small companies and learn their trade directly on the job. Others join companies that offer formal training

Figure 12.5 ■ *(Continued)*

programs. Some begin by getting experience in their technical specialty. Others earn a graduate degree beforehand. These options are compared below.

OPTION 1: ENTRY-LEVEL MARKETING WITH ON-THE-JOB TRAINING. Smaller manufacturers offer marketing positions in which people learn on the job. Elaine Carto, president of ABCO Electronics, believes small companies offer a unique opportunity; entry-level salespersons learn about all facets of an organization, and have a good possibility for rapid advancement (personal interview, February 10, 1999). Career counselor Phil Hawkins says, "It's all a matter of whether you prefer to be a big fish in a small pond or a small fish in a big pond" (personal interview, February 11, 1999).

Entry-level marketing offers immediate income and a chance for early promotion. A disadvantage, however, might be the loss of any technical edge one might have acquired in college.

OPTION 2: A MARKETING AND SALES TRAINING PROGRAM. Formal training programs offer the most popular entry into sales and marketing. Large to mid-size companies typically offer two formats: (a) a product-specific program, focused on a particular product or product line, or (b) a rotational program, in which trainees learn about an array of products and work in the various positions outlined in Figure 2. Programs last from weeks to months.

Former trainees Roger Cayer, of Allied Products, and Bill Collins, of Intrex, speak of the diversity and satisfaction such programs offer: specifically, solid preparation in all phases of marketing, diverse interaction with company personnel, and broad knowledge of various product lines (phone interviews, February 8, 1999).

Like direct entry, this option offers the advantage of immediate income and early promotion. With no chance to practice in their technical specialty, however, trainees might eventually find their technical expertise compromised.

OPTION 3: PRIOR EXPERIENCE IN ONE'S TECHNICAL SPECIALTY. Instead of directly entering marketing, some candidates first gain experience in their specialty. This option combines direct exposure to the workplace with the chance to sharpen technical skills in practical applications. In addition, some companies, such as Roger Cayer's, will offer marketing and sales positions to outstanding staff engineers, as a step toward upper management (phone interview, February 8, 1999).

Although this option delays a candidate's entry into technical marketing, industry experts consider direct workplace and technical experience key assets for career growth in any field. Also, work experience becomes an asset for applicants to top MBA programs (Shelley, 1997, pp. 30–31).

FIGURE 12.5 ■ *(Continued)*

OPTION 4: GRADUATE PROGRAM. Instead of direct entry, some people choose to pursue an MS degree in their specialty or an MBA. According to engineering professor Mary McClane, MS degrees are usually unnecessary for technical marketing unless the particular products are highly complex (personal interview, April 2, 1999).

In general, jobseekers with an MBA have a distinct competitive advantage. More significantly, new MBAs with a technical bachelor's degree and one to two years of experience command salaries from 10 to 30 percent higher than MBAs who lack work experience and a technical bachelor's degree. In fact, no more than 3 percent of job candidates offer a "techno-MBA" specialty, making this unique group highly desirable to employers (Shelley, 1997, p. 30).

A motivated student might combine graduate degrees. Dora Anson, president of Susimo Cosmic Systems, sees the MS/MBA combination as ideal preparation for technical marketing (1999).

One disadvantage of a full-time graduate program is lost salary, compounded by school expenses. These costs must be weighed against the prospect of promotion and monetary rewards later in one's career.

AN OVERALL COMPARISON BY RELATIVE ADVANTAGE. Table 1 compares the four entry options on the basis of three criteria: immediate income, rate of advancement, and long-term potential.

Table 1   Relative Advantages Among Four Technical-Marketing Entry Options

	Relative Advantages		
Option	Early, immediate income	Greatest advancement in marketing	Long-term potential
Entry level, no experience	yes	yes	no
Training program	yes	yes	no
Practical experience	yes	no	yes
Graduate program	no	no	yes

FIGURE 12.5 ■ *(Continued)*

CONCLUSION

Summary of Findings

Technical marketing and sales involves identifying, reaching, and selling the customer a product or service. Besides a solid technical background, the field requires motivation, communication skills, and interpersonal skills. This career offers job diversity and excellent income potential, balanced against hard work and relentless pressure to perform.

College graduates interested in this field confront four entry options: (1) direct entry with on-the-job training, (2) a formal training program, (3) prior experience in a technical specialty, and (4) graduate programs. Each option has benefits and drawbacks based on immediacy of income, rate of advancement, and long-term potential.

Interpretation of Findings

For graduates with a strong technical background and the right skills and motivation, technical marketing offers attractive career prospects. Anyone contemplating this field, however, needs to be able to enjoy customer contact and thrive in a highly competitive environment.

Those who decide that technical marketing is for them can choose among the various entry options:

- For hands-on experience, direct entry is the logical option.
- For sophisticated sales training, a formal program with a large company is best.
- For sharpening technical skills, prior work in one's specialty is invaluable.
- If immediate income is not vital, graduate school is an attractive option.

Recommendations

If your interests and abilities match the requirements, consider these suggestions:

1. To get a firsthand view, seek the advice and opinions of people in the field.
2. Before settling on an entry option, consider all its advantages and disadvantages and decide whether this option best coincides with your career goals. (Of course, you can always combine options during your professional life.)
3. When making any career decision, consider career counselor Phil Hawkins' advice: "Listen to your brain and your heart " (personal interview, February 11, 1999). Choose an option or options that offer not only professional advancement but also personal satisfaction.

Figure 12.5 ■ *(Continued)*

References

Anson, D. (1999, March 12). *Engineering graduates and the job market.* Lecture presented at the University of Massachusetts at Dartmouth.   *[lecture]*

Basta, N. (1984, September). Take a good look at sales engineering. *Graduating Engineer, 32,* 84–87.   *[journal article]*

College Placement Council. (1998). *CPC Annual* (42nd ed.). Bethlehem, PA: Author.   *[reference book—author/organization as publisher]*

Cornelius, H., & Lewis, W. (1983). *Career guide for sales and marketing* (2nd ed.). New York: Monarch Press.   *[book with two authors]*

Electronic sales positions. (1996). *The national job bank.* Holbrook, MA: Bob Adams, Inc.   *[directory entry—no author named]*

Jones, B. (1997, December). Giving women the business. *Harper's Magazine, 296* (1772), 47–58.

Occupational employment. (1997–98, Winter) *Occupational Outlook Quarterly, 41*(4), 6–24.   *[govt. periodical—no author named]*

Shelley, K. J. (1997, Fall). A portrait of the M.B.A. *Occupational Outlook Quarterly, 41*(3), 26–33.   *[govt. periodical—author named]*

Solomon, S. D. (1996, January). An engineer goes to Wall Street [10 pages]. *Technology Review* [Online serial], 99(i). Available: www.web.mit.edu/techreview/www   *[online article]*

Technology Marketing Group, Inc. (1998). *Services for clients* [Online Web site]. Available: www.technology-marketing.com   *[Web site]*

Tolland, M. (1999, April). *Alternate careers in marketing.* Presentation at Electro '99 Conference in Boston.   *[unpublished conference presentation]*

U.S. Department of Labor. (1997) *Tomorrow's jobs.* Washington, DC: Author.   *[govt. publication—no author named]*

Young, J. (1995, August). Can computers really boost sales? *Forbes ASAP,* 84–101.   *[magazine article—no vol. or issue number]*

Figure 12.5 ■ *(Continued)*

Typical components of long reports

After analyzing your audience and purpose, do some basic research. Then, sketch a rough outline with headings and subheadings for the report. The outline in Figure 12.6 is general enough that you can adapt it to your specific situation.

Most reports contain the following sections:

Abstract or executive summary. Reports are often read by many people: researchers, developers, managers, vice presidents, customers. For readers who are interested only in the big picture, the entire report may not be relevant, so most long reports are commonly preceded by an abstract (short) or an executive summary (longer). In this brief description, you explain the issue, describe how you researched it, and state your conclusion. Busy readers can then flip through the document to locate sections of importance to them.

I. Introduction
   A. Definition, Description, and Background
   B. Purpose of the Report, and Intended Audience
   C. Method of Inquiry
   D. Working Definitions (here or in a glossary)
   E. Limitations of the Study
   F. Scope of the Inquiry (topics listed in logical order)

II. Collected Data
   A. First Topic for Investigation
     1. Definition
     2. Findings
     3. Interpretation of findings
   B. Second Topic for Investigation
     1. First subtopic
       a. Definition
       b. Findings
       c. Interpretation of findings
     2. Second subtopic (and so on)

III. Conclusion
   A. Summary of Findings
   B. Overall Interpretation of Findings (as needed)
   C. Recommendations (as needed and appropriate)

FiGURE 12.6 ■ **A general outline for a report that you can adapt for your situation.**

Introduction. The introduction engages and orients the audience and provides background as briefly as possible for the given situation. Often, writers who are familiar with the product are tempted to write long introduction sections because they have a lot of background knowledge about the product or issue. But readers don't generally need long history lessons about the topic. In the introduction, identify the topic's origin and significance, define or describe the problem or issue, and explain the report's purpose. Briefly identify your research methods (interviews, literature searches, and so on). List working definitions, but if you have more than two or three, place definitions in a glossary. Finally, briefly state your conclusion. Don't make readers wade through the entire report to find out what you are recommending or advising.

Body. The body describes and explains your findings. Present a clear and detailed picture of the evidence, interpretations, and reasoning on which you will base your conclusion. Divide topics into subtopics, and use informative headings as aids to navigation. Figure 12.5 shows a long report that uses headings and subheadings in the body. The body of your report will vary greatly depending on the audience, topic, purpose, and situation.

Conclusion. The conclusion is important because it answers the original questions that sparked the analysis. In the conclusion, you summarize, interpret, and recommend. Although you have interpreted evidence at each stage of your analysis, your conclusion presents a broad interpretation and suggests a course of action where appropriate. Your conclusion should provide a clear and consistent perspective on the whole document. Don't introduce new ideas, facts, or statistics in the conclusion.

Appendices. Add one or more appendices to your report if you have large blocks of material or other documents that are relevant but will bog readers down if placed in the middle of the report itself. For example, if your report on the cost of electricity at your company refers to another report issued by the local utility company, you may wish to include this second report as an appendix.

Glossary. Use a glossary if your report contains more than two or three technical terms that may not be understood by all audience members. Use standard definitions in your glossary: Refer to company style guides or technical dictionaries.

Other items. A report may also include a letter of transmittal, title page, table of contents, list of figures, and Works Cited page. These items increase the report's usability, because they help readers find the information they are looking for. Look for examples of other reports written at your company or organization to see if there are any other required components.

Usability considerations

Clearly identify the problem or goal. To address the true purpose of the situation, you must carefully identify your goal. Begin by defining the main questions involved in the report and then outlining any subordinate questions. Your employer, for example, might pose this question: "Will a low-impact aerobics program significantly reduce stress among my employees?" Answering this question is the main goal of your report; however, this question leads to several others, such as, "What are the therapeutic claims for aerobics, and are these claims valid?" Create a goal statement, such as "The goal of this report is to examine and evaluate claims about the therapeutic benefits of low-impact aerobic exercise."

Provide enough, but not too much, information. Any usable analysis must address the needs, interests, and technical expertise of your audience. A long history of the development of the pacemaker may be interesting to you but inappropriate for your report. As you plan the report, find out how much of the information you've gathered readers need in order to make a decision. Also, make sure your technical terms are not too complex for your audience. If you have a mixed audience, provide a glossary where readers can look up unfamiliar terms. If your report is posted to a Web site, you can use hyperlinks for glossary terms.

Provide accurate information. Make sure your information is as accurate as possible and, to the best of your ability, without bias. Use reputable information sources, particularly for statistical data. Be careful when taking information from the Web; Web sites often sound credible but can be based on biased or inaccurate information (see Chapters 4 and 5). Also, make sure you interpret information fairly and provide valid conclusions based on your best research. Assume, for example, that you were writing a report to recommend the best brand of chainsaw for a logging company. In reviewing test reports, you learned that one brand, Bomarc, is easiest to operate but also has the fewest safety features. Both pieces of information should be included in the report, regardless of your personal preference for this brand.

Use appropriate visuals. As discussed in Chapter 8, visual information can make complex statistics and numerical data easy to understand. Graphs are especially useful for analyzing rising or falling trends, levels, and long-term forecasts. Tables and charts are helpful for comparing data. Photographs and diagrams are an excellent way to show a component or special feature. Be sure your visual is placed near the accompanying text, and be careful not to overuse visuals.

Use informative headings. Headings and subheadings in your report announce what each section contains. A heading "Data Analysis" does not really say much, whereas a heading "Physiological Effects and Health Risks" offers a clear, informative preview about the contents of a section.

Write clearly and concisely. Even readers who need every bit of information in your report don't want to be bogged down with prose that is cumbersome, long-winded, and hard to read. Keep your language crisp and clear. Use active voice whenever possible. Ask a colleague or editor to copyedit your report before it is printed.

Use action verbs. Especially when recommending a plan of action, use action verbs such as *examine, evaluate, determine,* or *recommend.* Avoid nominalizations: Don't use *determination* when you mean *determine,* for example.

Proposals

The situation

Proposals encourage an audience to take some form of direct action: to authorize a project, purchase a service or product, or otherwise support a specific plan for solving a problem. While they often contain the same basic elements as a report, proposals have one specific purpose: to propose an action or series of actions. This purpose differs from more generic reports, which can be used for other purposes (see previous section). Proposals can be called for in a variety of situations: a request to fund a training program for new employees; a suggestion to change the curriculum in your English or Biology department; a bid to the U.S. Defense Department on a missile contract. Depending on the situation, proposals may be short or long and may be written in the form of a report, a letter, or a memo.

Audience and purpose analysis

In science, business, industry, government, or education, proposals are written for any number of audiences: managers, executives, directors, clients, board members, or community leaders. Inside or outside the organization, these people review various proposals and then decide whether the plan is worthwhile, whether the project will materialize, and whether the service or product is useful. At the most general level, the purpose is to persuade your audience. More specifically, proposals often answer questions about the nature of the problem or product, the benefits of your proposed plan, cost, completion dates, schedules, and so on.

Types of proposals

Proposals may be solicited or unsolicited. Solicited proposals are those that have been requested by a client or customer. For example, if you represent an engineering firm specializing in highway construction, you may receive a request for proposal (RFP) from a local township asking you to bid on a road project. Typically, an RFP is issued to numerous companies, and your proposal will need to stack up against all the others. Unsolicited proposals are those that have not been specifically requested. If you are a new advertising agency in town, you may send out short proposals to local radio stations suggesting that they use your agency for their advertising needs.

Because the audience for a solicited proposal has made the request, you may not need to spend as much time introducing yourself or providing background on the product or service. For an unsolicited proposal (sometimes called a "cold call" in sales), you will need to catch your readers' attention quickly and provide incentives for them to continue reading: perhaps by printing a price comparison of your fees on the first page, for example.

Both solicited and unsolicited proposals can take the following forms:

Planning proposal. A planning proposal offers solutions to a problem or suggestions for improvement. It might be a request for funding to expand the campus newspaper, an architectural plan for new facilities at a ski area, or a plan to develop energy alternatives to fossil fuels. Figure 12.7 is a short planning proposal that was solicited and will be used internally within the company. The XYZ Corporation has contracted a team of communication consultants to design in-house writing workshops, and the consultants must convince the client that their methods will succeed. After briefly introducing the problem, the authors develop their proposal under two headings and several subheadings, making the document easy to read and to the point.

Research proposal. Research (or grant) proposals request approval and funding for research projects. A chemist at a university might address a research proposal to the Environmental Protection Agency for funds to identify toxic contaminants in local groundwater. Research proposals are solicited by many agencies: the National Science Foundation, the National Institutes of Health, and others. Each agency has its own requirements and guidelines for proposal format and content. Successful research proposals follow these guidelines, and carefully articulate the goals of the project. In these cases, proposal readers will generally be other scientists, so writers can use language that is appropriate for other experts. Other research proposals might be submitted by students requesting funds for undergraduate research projects. In Figure 12.8, the writer used the prescribed form and was careful to relate his research to the goals of the program.

Sales proposal. Sales proposals offer services or products and may be solicited or unsolicited. If solicited, several firms may be competing for the contract, so

Dear Mary:

Thanks for sending the writing samples from your technical support staff. Here is what we're doing to design a realistic approach.

Assessment of Needs
After conferring with technicians in both Jack's and Terry's groups and analyzing their writing samples, we identified this hierarchy of needs:

- improving readability
- achieving precise diction
- summarizing information
- organizing a set of procedures
- formulating various memo reports
- analyzing audiences for upward communication
- writing persuasive bids for transfer or promotion
- writing persuasive suggestions

Proposed Plan
Based on the needs listed above, we have limited our instruction package to eight carefully selected and readily achievable goals.

Course Outline. Our eight, two-hour sessions are structured as follows:

1. achieving sentence clarity
2. achieving sentence conciseness
3. achieving fluency and precise diction
4. writing summaries and abstracts
5. outlining manuals and procedures
6. editing manuals and procedures
7. designing various reports for various purposes
8. analyzing the audience and writing persuasively

Classroom Format. The first three meetings will be lecture-intensive with weekly exercises to be done at home and edited collectively in class. The remaining five weeks will combine lecture and exercises with group editing of work-related documents. We plan to remain flexible so we can respond to needs that arise.

Limitations
Given our limited contact time, we cannot realistically expect to turn out a batch of polished communicators. By the end of the course, however, our students will have begun to appreciate writing as a deliberate process.

If you have any suggestions for refining this plan, please let us know.

Figure 12.7 ■ Planning proposal.

To: Dr. John Lannon
From: T. Sorrells Dewoody
Date: March 16, 20XX
Subject: *Proposal for Determining the Feasibility of Marketing Dead Western White Pine*

Introduction
Over the past four decades, huge losses of western white pine have occurred in the northern Rockies, primarily attributable to white pine blister rust and the attack of the mountain pine beetle. Estimated annual mortality is 318 million board feet. Because of the low natural resistance of white pine to blister rust, this high mortality rate is expected to continue indefinitely.

If white pine is not harvested while the tree is dying or soon after death, the wood begins to dry and check (warp and crack). The sapwood is discolored by blue stain, a fungus carried by the mountain pine beetle. If the white pine continues to stand after death, heart cracks develop. These factors work together to cause degradation of the lumber and consequent loss in value.

Statement of Problem
White pine mortality reduces the value of white pine stumpage because the commercial lumber market will not accept it. The major implications of this problem are two: first, in the face of rising demand for wood, vast amounts of timber lie unused; second, dead trees are left to accumulate in the woods, where they are rapidly becoming a major fire hazard here in northern Idaho and elsewhere.

Proposed Solution
One possible solution to the problem of white pine mortality and waste is to search for markets other than the conventional lumber market. The last few years have seen a burst of popularity and growing demand for weathered barn boards and wormy pine for interior paneling. Some firms around the country are marketing defective wood as specialty products. (These firms call the wood from which their products come "distressed," a term I will use hereafter to refer to dead and defective white pine.) Distressed white pine quite possibly will find a place in such a market.

Scope
To assess the feasibility of developing a market for distressed white pine, I plan to pursue six areas of inquiry:

1. What products presently are being produced from dead wood, and what are the approximate costs of production?

FIGURE 12.8 ■ Undergraduate research proposal.

2. How large is the demand for distressed-wood products?
3. Can distressed white pine meet this demand as well as other species meet it?
4. Does the market contain room for distressed white pine?
5. What are the costs of retrieving and milling distressed white pine?
6. What prices for the products can the market bear?

Methods
My primary data sources will include consultations with Dr. James Hill, Professor of Wood Utilization, and Dr. Sven Bergman, Forest Economist—both members of the College of Forestry, Wildlife, and Range. I will also inspect decks of dead white pine at several locations, and visit a processing mill to evaluate it as a possible base of operations. I will round out my primary research with a letter and telephone survey of processors and wholesalers of distressed material.

Secondary sources will include publications on the uses of dead timber, and a review of a study by Dr. Hill on the uses of dead white pine.

My Qualifications
I have been following Dr. Hill's study on dead white pine for two years. In June of this year I will receive my B.S. in forest management. I am familiar with wood milling processes and have firsthand experience at logging. My association with Drs. Hill and Bergman gives me the opportunity for an in-depth feasibility study.

Conclusion
Clearly, action is needed to reduce the vast accumulations of dead white pine in our forests. The land on which they stand is among the most productive forests in northern Idaho. By addressing the six areas of inquiry mentioned earlier, I can determine the feasibility of directing capital and labor to the production of distressed white pine products. With your approval I will begin research at once.

FIGURE 12.8 ■ *(Continued)*

your proposal may be ranked by a committee against several others. Sales proposals can be cast as letters if the situation calls for them to be brief. If the situation requires a longer proposal, you follow the guidelines for writing a report: Use a title page, make sure you have clear headings, and so on. A successful sales proposal persuades customers that your product or service surpasses those of any competitors. In the sample sales proposal in Figure 12.9, the writer explains why his machinery is best for the job, what qualifications his company can offer, and what costs are involved. What you include in a sales proposal is determined by the guidelines from the client or by a thorough analysis of what kinds of information your audience needs.

Typical components of proposals

After conducting your audience and purpose analysis, you should perform some basic research. For example, you might look into the very latest technology for solving the problem or doing the project; compare the costs, benefits, and drawbacks of various approaches; contact others in your field for their suggestions; and so on. Then, generate a rough outline with headings and subheadings for the proposal. The outline in Figure 12.10 is general enough to adapt to your specific situation.

As noted, proposals can be short (letter or memo format) or long (report format). For a long proposal, include the components and supplements ordinarily contained in a long report: abstract, introduction, body, conclusion, and appendices. Include a letter of transmittal, especially if your proposal is unsolicited.

Background. A background section (sometimes used as an introduction) can be brief or long. In Figure 12.7, the writer's opening sentence ("Thanks for sending the writing samples from your technical support staff.") provides a quick reminder of the context for the project. This sentence is brief because the writer correctly assumes that the reader is very familiar with the project. For a new audience, this single sentence might need to be expanded into a longer paragraph of background on the project. The background section may contain a statement of the problem or issue. If the topic warrants it, the background section may take up several pages.

Objective. If your audience needs this information spelled out, you may wish to provide a clear statement of the proposal's objectives: "Our objective is to offer a plan to make areas of the library quiet enough for serious study."

Clear statement of what is being proposed. Whether your proposal is short or long, make it easy for your audience to locate the exact details of what you are proposing. In the research proposal shown in Figure 12.8, the third heading, "Proposed Solution" is the obvious place for readers to turn if they want to see the details of the solution.

Subject: *Proposal to Dig a Trench and Move Boulders at Bliss Site*

Dear Mr. Haver:

I've inspected your property and would be happy to undertake the land-scaping project necessary for the development of your farm.

The backhoe I use cuts a span 3 feet wide and can dig as deep as 18 feet—more than an adequate depth for the mainline pipe you wish to lay. Because this backhoe is on tracks rather than tires, and is hydraulically operated, it is particularly efficient in moving rocks. I have more than twelve years of experience with backhoe work and have completed many jobs similar to this one.

After examining the huge boulders that block access to your property, I am convinced they can be moved only if I dig out underneath and exert upward pressure with the hydraulic ram while you push forward on the boulders with your D-9 Caterpillar. With this method, we can move enough rock to enable you to farm that now inaccessible tract. Because of its power, my larger backhoe will save you both time and money in the long run.

This job should take 12 to 15 hours, unless we encounter subsurface ledge formations. My fee is $200 per hour. The fact that I provide my own dynamiting crew at no extra charge should be an advantage to you because you have so much rock to be moved.

Please phone me anytime for more information. I'm sure we can do the job economically and efficiently.

FiɢuRE **12.9** ■ **Sales proposal.**

Budget and costs. If your proposal involves financial costs, make sure your cost and budget section is accurate and easy to understand. If you work with an accountant or other financial specialist, ask that person to check your figures. If the proposal is solicited, make sure you follow the client's guidelines for creating a budget.

Usability considerations

Understand the audience's needs. The proposal audience wants specific suggestions to meet their specific needs. Their biggest question is "What will this plan do for me?" Make your proposal demonstrate a clear understanding of the client's problem and expectations, and then offer an appropriate solution. In Figure 12.7 (planning proposal), the writer begins with a clear assessment of needs, and then moves quickly into a proposed plan of action.

I. **Introduction**
   A. Statement of Problem and Objective
   B. Background
   C. Need
   D. Benefits
   E. Qualifications of Personnel
   F. Data Sources
   G. Limitations and Contingencies
   H. Scope

II. **Body**
   A. Methods
   B. Timetable
   C. Materials and Equipment
   D. Personnel
   E. Available Facilities
   F. Needed Facilities
   G. Cost
   H. Expected Results
   I. Feasibility

III. **Conclusion**
   A. Summary of Key Points
   B. Request for Action

FIGURE 12.10 ■ A general proposal outline that you can adapt to your situation.

Maintain a clear focus on benefits. Show your audience that you understand what they will gain by adopting your plan. The planning proposal in Figure 12.7 includes a numbered list of exactly what tasks will be accomplished after the technical support staff takes the instruction courses.

Use honest and supportable claims. Because they typically involve large sums of money as well as contractual obligations, proposals require a solid ethical and legal foundation. False promises not only damage the writer's or company's reputation, they also invite lawsuits. If the solutions you offer have certain limitations, make sure you say so. For example, if you are proposing to install a new network server, make it clear what capabilities this server has, as well as what it cannot do under certain circumstances.

Use appropriate visuals. See Chapter 9 for discussion of visuals.

Write clearly and concisely. Make sure your document is easy to read, uses action verbs, and avoids puffed up language or terms that are too technical for your audience. If necessary for a mixed audience with differing technical levels, include a glossary.

Use convincing language. There is no need to be coy when writing a proposal. You are trying to sell yourself or your ideas. So be sure to write a document that will move people to action. Use statistics ("for the third year in a row, our firm has been ranked as the #1 architecture firm in the Midwest.") and direct sentences ("We know you will be satisfied with the results.").

Review Checklist

Type of Communication	Is Called for When You Need to ...	Specific Types
Definitions and descriptions	*define*, or explain, a term or concept that is specialized or unfamiliar to an audience	Brief Expanded Specifications
Long reports	inform an audience, offer a solution to a problem, report progress, or make a recommendation in a thorough manner	Causal Comparative Feasibility
Proposals	encourage an audience to take some form of direct action: to authorize a project, purchase a service or product, or otherwise support a specific plan for solving a problem	Planning Research Sales

Exercises

1. FOCUS ON ⟶WRITING. Choose a situation and an audience, and prepare an expanded technical definition specifically designed for this audience's level of technical understanding. In addition to the usability considerations on pages 256–258, use these guidelines:
   a. *Decide on the level of detail.* Definitions vary greatly in length and detail, from a few words in parentheses to a complete essay. How much does this audience need in order to follow your explanation or grasp your point?
   b. *Classify the term precisely.* The narrower your class, the clearer your meaning. Stress is classified as an applied force; saying that stress "is what . . ." or "takes place when . . ." fails to reflect a specific classification. Diabetes is precisely classified as a metabolic disease, not as a medical term.

c. *Differentiate the term accurately.* If the distinguishing features are too broad, they will apply to more than this one item. A definition of brief as a "legal document used in court" fails to differentiate a brief from all other legal documents (wills, affidavits, and so on).

d. *Avoid circular definitions.* Do not repeat, as part of the distinguishing feature, the word you are defining. "Stress is an applied force that places stress on a body" is a circular definition.

e. *Expand your definition selectively.* Begin with a sentence definition and select from a combination of development strategies.

f. *Use negation to show what a term does not mean.* For example: Raw data is not "information"; data only becomes information after it has been evaluated, interpreted, and applied.

2. **Focus on ⮕Writing.** Choose another situation and audience, and prepare a technical description for this audience's level of technical understanding. As you prepare your description, refer to the usability considerations on pages 256–258 and to these guidelines:

a. *Always begin with some type of orienting statement.* Descriptions rarely call for a standard topic or thesis statement, because their goal is to simply catalog the details that readers can visualize. Any description, however, should begin by telling readers what to look for.

b. *Choose descriptive details to suit your purpose and the reader's needs.* Select only those details that advance your meaning. Use objective details to provide a picture of something exactly as a camera would record it. Use subjective details to convey your impressions—to give readers a new way of seeing or appreciating something, as in a marketing brochure (p. 240).

c. *Select details that are concrete and specific enough to convey an unmistakable picture.* Most often, description works best at the lowest levels of abstraction and generality.

Vague	*Exact*
at high speed	80 miles per hour
a tiny office	an 8' x 10' office
the seal	the rubber O-ring

d. *Use sensory details as needed.* Allow readers to see, hear, and feel. Let readers touch and taste. Use vivid comparisons to make the picture come to life. Rely on action verbs to convey the energy of movement.

e. *Order details in a clear sequence.* Descriptions generally follow a spatial or general-to-specific order—whichever parallels the angle of vision readers would have if viewing the item. Details may also be arranged according to the dominant impression desired.

3. **Focus on ⮕Writing.** Choose a specific situation and audience, and prepare a long report that documents a causal, comparative, or feasibility

analysis—or some combination of these types. As you prepare your report, refer to the usability considerations on pages 271–272 and observe the following guidelines:

For Causal Analysis

a. *Be sure the cause fits the effect.* Keep in mind that faulty causal reasoning is extremely common, especially when people ignore other possible causes or confuse mere coincidence with causation.

b. *Make the links between cause and effect clear.* Identify the immediate cause (the one most closely related to the effect) as well as the distant causes (the ones that precede the immediate cause). For example, the immediate cause of a particular airplane crash might be a fuel tank explosion, caused by a short circuit in frayed wiring, caused by faulty design or poor quality control by the manufacturer. Discussing only the immediate cause often only scratches the surface of the problem.

c. *Clearly distinguish among possible, probable, and definite causes.* Unless the cause is obvious, limit your assertions by using *perhaps, probably, maybe, most likely, could, seems to, appears to,* or similar qualifiers that prevent you from making an insupportable claim.

For Comparative Analysis

a. *Rest the comparison on a clear and definite basis.* Make comparisons on the basis of costs, uses, benefits/drawbacks, appearance, or results. In evaluating the merits of competing items, identify your specific criteria and rank them in order of importance.

b. *Give both items balanced treatment.* Discuss points for each item in identical order.

c. *Support and clarify the comparison or contrast through credible examples.* Use research, if necessary, to find examples that readers can visualize.

d. *Follow either a block pattern or a point-by-point pattern.* In the block pattern, first one item is discussed fully, then the next. Choose a block pattern when the overall picture is more important than the individual points.

   In the point-by-point pattern, one point about both items is discussed, then the next point, and so on. Choose a point-by-point pattern when specific points might be hard to remember unless placed side by side.

Block pattern	**Point-by-point pattern**
Item A	first point of A/first point of B, etc.
first point	
second point	
third point, etc.	

Block pattern	**Point-by-point pattern**
Item B	second point of A/second point of B, etc.
first point	
second point	
third point, etc.	

e. *Order your points for greatest emphasis.* Try ordering your points from least to most important, dramatic, useful, or reasonable. Placing the most striking point last emphasizes it best.

f. *If you are writing an evaluative comparison, offer your final judgment.* Base your judgment squarely on the criteria presented.

For Feasibility Analysis

a. *Consider the strength of supporting reasons.* Decide carefully which are the best reasons supporting the action or decision being considered, based on solid evidence.

b. *Consider the strength of opposing reasons.* Remember that people usually see only what they want to see. Avoid the temptation to overlook or downplay opposing reasons, especially for an action or decision that you have been promoting. Consider alternate points of view and examine all the evidence.

c. *Recommend a realistic course of action.* After weighing all the pros and cons, make your recommendation—but be prepared to backtrack if you discover that what seemed like the right course of action turns out to be wrong.

1. Focus on ➠Writing. Develop an unsolicited proposal for solving a problem, improving a situation or satisfying a need in your school, community, or job. Address a clearly identified audience of decision makers. As you prepare your proposal, refer to the usability considerations on pages 278–280 and to these guidelines:

   a. *Spell out the problem (and its causes) clearly and convincingly.* Give enough detail for your audience to appreciate the problem's importance. Answer this implied question: "Why is this such a big deal?"

   b. *Point out the benefits of solving the problem.* Answer this implied question: "Why should we spend the time, money, and effort to do this?"

   c. *Offer a realistic solution.* Stick to claims or assertions you can support. Answer this implied question: "How do we know this will work?"

   d. *Address anticipated objections to your solution.* Consider carefully the audience's skepticism on this issue. Answer this implied question: "Why should we accept the things that seem wrong with your plan?"

   e. *Induce readers to act.* Decide exactly what you want readers to do and give reasons why they should be the ones to act. Answer this implied question: "What action am I supposed to take?"

The Collaboration Window

Work with three or four others in your major and discuss proposals in your profession. When might you need to write a proposal (purpose) and for whom (audience)? Discuss whether the proposal would be solicited or unsolicited and which type you might need to write (planning, research, sales). Create a rough outline for a possible proposal. As a group, report your findings to class and in a memo to your instructor.

The Global Window

Compare the sort of proposals regularly done in the United States with those created in other countries. For example, is the format the same? Are there more proposals for certain purposes in the United States than in another country? You can learn about this topic by interviewing an expert in international business (someone you meet on the job, during an internship, or through your advisor). You can also search the Web for information on international business communication. Describe your findings in a short memo that you will share with your classmates.

Click on This

The Internet makes it easy for audiences from anywhere to access scientific and technical definitions. For example, the U.S. National Institutes of Health (NIH) has created a Web site with definitions of rare diseases, located at

rarediseases.info.nih.gov/ord/glossary_a-e.html

A similar example is the Students' Cloud Observations Online site (S'COOL), sponsored by NASA. It offers simple definitions of weather terms for school-age children, set up as a glossary. Go to

asd-www.larc.nasa.gov/SCOOL

and click on glossary.

For a long report sample, see the comparative analysis of gender equity in education at

nces.ed.gov/pubs2000/200030.pdf

Appendix **A**

Grammar

GRAMMAR ISSUES

PUNCTUATION

LISTS

AVOIDING SENTENCE FRAGMENTS AND RUN-ON SENTENCES

USAGE (COMMONLY MISUSED WORDS)

SUBJECT-VERB AND PRONOUN-ANTECEDENT-INTERCEDENT AGREEMENT

FAULTY MODIFICATION

MECHANICS

GRAMMAR TOPIC 1: PUNCTUATION

Punctuation Mark	How it's used	Examples
Period (.) For more on the period, see sentence fragments and run-on sentences on pages 290, 292.	A period is used at the end of a complete idea to signal the end of a sentence.	*Use of white space is an important consideration when designing documents. However, there are no strict rules about how much white space should be on a page.*
Semicolon (;) For more on the semicolon, see sentence fragments and run-on sentences on pages 290, 292.	A semicolon is used at the end of a complete idea to signal a pause, but not the end of the sentence. A semicolon is always followed by another complete idea that is closely related to the one before it.	*Use of white space is an important consideration when designing documents; however, there are no strict rules about how much white space should be on a page.*
	A semicolon can also be used between items in a series when the items themselves contain commas.	*This summer we plan to visit relatives in Milwaukee, Wisconsin; Indianapolis, Indiana; and Des Moines, Iowa.*
Colon (:) For more on the colon, see lists on page 288.	Like the semicolon, a colon is used at the end of a complete idea to signal a pause. However, a colon doesn't need to be followed by another complete idea. Usually, the information that follows the colon explains or clarifies the information that comes before the colon. Often a colon is used *after a complete sentence* to introduce a list.	*We can't reopen for business this week: The renovation work is not yet complete.* *Before we can reopen for business, we will need to order the following supplies: printer paper, printer cartridges, envelopes, and postage.*
	A colon is not used to introduce a list after "including" or "such as."	*We will need to order several office supplies, including printer paper, printer cartridges, envelopes, and postage.*
Comma (,) For more on the comma, see sentence fragments and run-on sentences on pages 290–292.	A comma signals a pause within a sentence.	*Sue Jones, the first person we interviewed, was the most qualified candidate.*

(continued)

A comma can be used between two complete ideas if it's used with a coordinating conjunction (and, but, or, nor, yet).	*We interviewed Sue Jones first, and she was the most qualified candidate.*
A comma is used to separate an incomplete idea from a complete idea when the incomplete idea comes first.	*Because she has the most experience, we all agreed that Sue Jones was the most qualified candidate.*
When the incomplete idea comes after the complete idea, no comma should be used.	*We all agreed that Sue Jones was the most qualified candidate because she has the most experience.*

Apostrophe (')

Use an apostrophe to form possessives. For words that do not end in *s*, use an apostrophe followed by *s* to form the possessive. For words that do end in *s*, use only the apostrophe.	*When you design a document, the user's needs should always come first.* *When you design documents, users' needs should always come first.*
Also use apostrophes to form conjunctions.	*Can't, shouldn't, wouldn't, won't, etc.*

Quotation Marks ("")

Use quotation marks around words that you treat as terms.	*Most survey respondents answered "no" to the first question.*
NOTE: Periods and commas belong inside quotation marks. Colons and semicolons belong outside quotation marks. Question marks belong inside quotation marks if they are part of the material being quoted. Otherwise they belong outside.	

Italics/Underlining

Use italics or underlining for titles of books, journals, magazines, films, and newspapers. Also use them for foreign words or technical terms and for special emphasis.	*Star Wars is still my favorite movie.* *Do not leave your child unsupervised in the infant swing.*

(continued)

Punctuation Mark	How it's used	Examples
Parentheses ( )		
	Use parentheses to set off explanatory material that could be deleted without altering the meaning of the sentence.	*In general, the survey indicates that people are satisfied with the current document (see Appendix B for complete survey results).*
Dashes (—)		
	Use a dash to set off explanatory material that you want to emphasize.	*Storyteller—a collection of short stories by Leslie Marmon Silko—is an excellent book for use in introductory literature classes.*

GRAMMAR TOPIC 2: LISTS

Embedded lists

An embedded list is a series of items integrated into a sentence, as in the example below:

> The file menu allows you to perform the following tasks: create a new file, open an existing file, and save a current file.

Embedded lists can be numbered, as in the example below. To number an embedded list, use parentheses around the numerals and either commas or semicolons between the items.

> If you wish to apply for admission to the program, you must (1) submit official transcripts from all previous academic work, (2) complete a department application form, (3) submit at least three letters of recommendation from former employers or teachers, and (4) submit GRE scores.

Vertical lists

Embedded lists are appropriate when you only need to list a few short items. Vertical lists, with numerals, letters, or bullets, are more appropriate than embedded lists when you need to list several items. Use numerals or letters if the items in the list belong in a particular order (steps in a procedure, for example) and bullets if the order of items is unimportant. Listing items vertically is also a way to increase the white space on a page and to draw a reader's attention to the contents of the list. The examples below illustrate how to introduce and punctuate vertical lists.

You can introduce a vertical list with a sentence that contains "the following" or "as follows." End the introductory sentence with a colon.

> Before the second week of class, students must purchase *the following*:
>
> • Printer card
> • Three-ring notebook
> • Course packet
> • Writing handbook

You can also introduce a vertical list with a sentence that ends with a noun. Again, end the introductory sentence with a colon.

> Before the second week of class, students must purchase *four items:*
>
> 1. Printer card
> 2. Three-ring notebook
> 3. Course packet
> 4. Writing handbook

You can also introduce a vertical list with a sentence that is not grammatically complete without the list items.

> To adjust the volume of sound input devices:
>
> 1. Choose Control Panels from the main menu.
> 2. Open the Sound Control panel.
> 3. Select Volumes from the pop-up menu.
> 4. Drag the sliders up and down to achieve the desired volume.

DO NOT use a colon to introduce a list with a sentence that ends with a verb, a preposition, or an infinitive. Either remove the colon or revise the sentence.

INCORRECT because the introductory sentence ends with verb.

> Before the second week of class, students *need:*
>
> • A printer card
> • An activated university email account
> • A course packet

INCORRECT because the introductory sentence ends with a preposition.

> Before the second week of class, students need *to:*
>
> • Purchase a printer card
> • Activate a university email account
> • Purchase a course packet

INCORRECT because the introductory sentence ends with an infinitive.

> Before the second week of class, students need *to buy:*
>
> * A printer card
> * A three-ring notebook
> * A course packet
> * A writing handbook

If another sentence comes after the sentence that introduces a list, use periods after both sentences. **DO NOT** use a colon to introduce the list.

> The next step is to configure the following fields. Consult Chapter 3 for more information on each field.
>
> * Serial port
> * Baud rate
> * Data bits
> * Stop bits

Note that some of the above examples use a period after each list item and some do not. As a general rule, use a period after each list item if any of the items contains a complete sentence. Do not use a period if none of the list items contains a complete sentence. Also note that items included in a list should be grammatically parallel and comparable. For more on parallelism, see page 40.

Grammar topic 3: Avoiding sentence fragments and run-on sentences

Avoiding sentence fragments

A sentence fragment is a grammatically incomplete sentence. Writers can avoid sentence fragments by understanding what it means for a sentence to be grammatically complete. A grammatically complete sentence consists of at least one subject-verb combination and expresses a complete thought. It might include more than one subject-verb combination and it might include other words or phrases as well. All the examples below are grammatically complete sentences because they all contain at least one subject-verb combination, and they all express complete thoughts.

> This book summarizes recent criminal psychology research.
>   Subject   Verb
>
> The smudge tool creates soft effects.
>   Subject Verb
>
> The table of contents is still incomplete.
>   Subject   Verb

A sentence fragment might contain a subject-verb combination, but it doesn't express a complete idea. The example below is a sentence fragment. Even though it contains a subject-verb combination, it doesn't express a complete thought.

> Although the report was not yet complete.
>   Subject   Verb

This wording leaves a reader waiting for something to complete the thought. To make the sentence grammatically complete, add another subject-verb combination that completes the thought.

> Although the report was not yet complete, I began editing.
>   Subject   Verb   Subject   Verb

Watch out for *although* and other words like it, including *because, if, while, since, when,* and *unless.* These are called subordinating conjunctions: When any of these words is combined with a subject-verb combination, it produces a subordinate clause (a clause that expresses an incomplete idea). This incomplete idea can only be turned into a complete sentence if another subject-verb combination that does express a complete idea is added on.

There are various other kinds of sentence fragments as well. For example, the group of words below is a sentence fragment because it contains no verb.

> DesignPro, a brand new desktop publishing program.

It can be turned into a complete sentence in a couple of different ways.

> DesignPro, a brand new desktop publishing program, will be available soon.

or

> DesignPro is a brand new desktop publishing program.

Watch out for sentences that seem to contain a subject-verb combination but actually do not. Verb forms that end in -*ing,* such as *being* in the phrase below, are not complete verbs.

> Dale being a document design expert.

This fragment can be turned into a complete sentence by substituting the verb *is* for the verb *being.*

> Dale is a document design expert.

Avoiding run-on sentences

Whereas a sentence fragment is a grammatically incomplete sentence, a run-on sentence suffers from the opposite problem: It contains too many grammatically complete sentences joined together as one. For example, the run-on sentence below contains two subject-verb combinations, each of which expresses a complete thought:

> For emergencies, we dial 911 for other questions, we dial 088.
>   Subject Verb   Subject Verb

This sentence can be repaired in various ways. One possibility is to divide it into two sentences.

> For emergencies, we dial 911. For other questions, we dial 088.

Another possibility is to use a semicolon to join the two parts of the sentence. This option indicates a break that is not quite as strong as the period and, therefore, signals to the reader that the two items are closely related.

> For emergencies, we dial 911; for other questions, we dial 088.

Another possibility is to add a coordinating conjunction (*and, but, or, yet, nor*).

> For emergencies, we dial 911, but for other questions, we dial 088.

One **unacceptable** repair would be to add a comma. Doing so would produce a comma splice, which is another kind of run-on sentence.

> For emergencies, we dial 911, for other questions, we dial 088.

Another **unacceptable** repair would be to add a transitional word such as *however, consequently,* or *therefore.* These words alone or with a comma are not appropriate for joining two complete sentences.

> For emergencies, we dial 911, however, for other questions, we dial 088.

Instead, words such as *however, consequently,* and *therefore* (conjunctive adverbs) must be used with a semicolon or period.

> For emergencies, we dial 911; however, for other questions, we dial 088.
>
> For emergencies, we dial 911. However, for other questions, we dial 088.

GRAMMAR TOPIC 4: USAGE (COMMONLY MISUSED WORDS)

Be aware of the following pairs of words, which are commonly confused with each other in speech and in writing.

Word Pair	Examples
Affect means "to influence."	Sleep deprivation negatively *affects* driving ability.
Effect used as a noun means "a result."	Sleep deprivation has a negative *effect* on driving ability.
Effect used as a verb means "to cause" or "to bring about."	Management believes that the new policy will *effect* an increase in productivity.
Among is used with three or more people or items.	We can now divide the work *among* the three of us.
Between is used with two people or items.	We used to have to divide all of our work *between* the two of us.
Continual means "repeatedly."	Most young children need to be *continually* reminded to brush their teeth after meals.
Continuous means "nonstop."	After three hours in the car, the *continuous* static on the radio got on my nerves.
Disinterested means "neutral" or "objective."	Scientists are expected to be *disinterested* in their subject matter.
Uninterested means "bored" or "unconcerned."	Most of the students were *uninterested* in memorizing the grammar rules they were supposed to know for the quiz.
Farther refers to physical distances that can be measured.	I now live *farther* from the grocery store than I used to.
Further refers to abstract distances that can't be measured.	If you have any *further* questions, you'll need to speak with a supervisor.
Fewer refers to quantities that can be counted.	Most students like to take *fewer* credit hours in the spring quarter than they do in the winter quarter.
Less refers to quantities that cannot be counted.	The teacher was *less* concerned than usual about attendance this week.
Infer means "to guess" or "to speculate."	Several people who heard the victim's story were able to *infer* who committed the crime.
Imply means "to insinuate."	Her remarks *implied* that I was at fault.
Lay means to "put down" or "set down." It requires a direct object.	Please *lay* all your books on the floor during the exam.
Lie means "to recline." It does not require a direct object. (Note that the past tense of *lie* is *lay.*)	Dentists usually advise their patients to *lie* down for several hours after any kind of dental work that requires anesthesia.

(continued)

Word Pair	Examples
Like should be followed by a noun, not by a subject-verb combination.	You look *like* a million bucks.
As if should be followed by a subject-verb combination.	You look *as if* you could use a couple more hours of sleep.
Percent should be used with a specific number to represent a figure.	Only two *percent* of survey respondents approve of the current system.
Percentage should be used to refer to an unspecified amount.	Only a small *percentage* of survey respondents approve of the current system.
Principle means "a fundamental rule or guideline."	Sometimes employees must choose whether to act in accordance with their own personal *principles* or those of the corporation.
Principal, when used as a noun, means "the chief person."	The school *principal* had to call several parents to let them know about the incident.
Principal, when used as an adjective, means "chief."	Her *principal* goal in seeking a new job is to earn more money.

Grammar topic 5: Subject-verb and pronoun-antecedent agreement

Subject–verb agreement

(1) Make the verb agree with its subject, not with a word that comes between.

> Examples   The *tulips* in the pot on the balcony *need* watering.
>
> The *teacher,* as well as his assistant, *was* reprimanded for his behavior.

(2) Treat compound subjects connected by *and* as plural.

> Example   *Terry and Julie enjoy* collaborating on writing projects.

(3) With compound subjects connected by *or* or *nor,* make the verb agree with the part of the subject nearer to the verb.

> Examples   If my mother or *sister calls,* tell her I will be right back.
>
> Neither the professor nor *the students were* able to figure out what was going on.

(4) Treat most indefinite pronouns (*anybody, each, everybody, etc.*) as singular.

> EXAMPLE   Almost *everybody* who registered for the class *was* there on the first day.

(5) Treat collective nouns as singular unless the meaning is clearly plural.

> EXAMPLES   The *group respects* their leader.
>
> The *team were* debating among themselves.

(6) Make the verb agree with its subject even when the subject follows the verb.

> EXAMPLE   At the back of the room *were* a *stereo and a small chair.*

Pronoun-antecedent agreement

(7) Make pronouns and the words they refer to agree with each other.

> EXAMPLES   *Everyone* should proceed at *his or her* own pace.
>
> The *committee* finally decided to proceed with *its* building plan.
>
> The *committee* put *their* signatures on the final draft of the report.

Grammar topic 6: Faulty modification

Faulty modification occurs either when a phrase has no word to modify, or when the position of a phrase within a sentence makes it difficult to determine which word the phrase is supposed to modify. For example, the sentence below makes no sense because the phrase "taking a shower" has no word to modify: This phrase is a *dangling modifier.*

I Taking a shower, the baby crawled out of his crib.

Revised as follows, the meaning of the sentence is much more clear:

I While I was taking a shower, the baby crawled out of his crib.

Additional examples of sentences containing dangling modifiers appear below.

Faulty	*Dialing the phone,* the cat ran out the open door.
Revised	As Joe dialed the phone, the cat ran out the open door.

Faulty	*After completing the student financial aid application form,* the Financial Aid Office will forward it to the appropriate state agency.
Revised	After you complete the student financial aid application form, the Financial Aid Office will forward it to the appropriate state agency.

The sentence below is unclear because the phrase "that was really boring" is supposed to modify "a weekend." Because of the position of this phrase in the sentence, a reader would mistakenly think it was the cabin, rather than the weekend, that was boring: This phrase is a *misplaced modifier.*

▌ We spent a weekend at the cabin that was really boring.

Revised as follows, the meaning of the sentence is much more clear.

▌ We spent a really boring weekend at the cabin.

Additional examples of sentences containing misplaced modifiers appear below.

Faulty	Joe typed another memo on our computer *that was useless.*
Revised	Joe typed another useless memo on our computer.

Faulty	He read a report on the use of nonchemical pesticides *in our conference room.*
Revised	In our conference room, he read a report on the use of non-chemical pesticides.

Faulty	She volunteered to deliver the radioactive shipment *immediately.*
Revised	She immediately volunteered to deliver the radioactive shipment.

Parallel structure

Parallel structure is a fancy way of saying that similar items should be expressed in similar grammatical form. For example, the following sentence is not parallel.

▌ She enjoys many outdoor activities, including *running, kayaking,* and *the design of new wilderness trails.*

This sentence is essentially a list of items. The first two items, *running* and *kayaking,* are expressed as verbs with *-ing* endings. The third item, *the design of* *new wilderness trails,* is not a verb. It is a nominalization. To make this sentence parallel, you would revise as follows:

> She enjoys many outdoor activities, including *running, kayaking,* and *designing new wilderness trails.*

Also make items in bulleted and numbered lists parallel. (For more on bulleted and numbered lists, see pages 288–290.) For example, in the list included in the paragraph below, the first two items are parallel, but the third is not.

> As the enclosed résumé illustrates, I have held several jobs relevant to the security management position your company is currently attempting to fill:
>
> • Park ranger for the City of Minneapolis
> • Security guard at the Mall of America
> • I also worked as a part-time bouncer at Pete's Tavern, a local bar

To make the third item parallel with the first two, revise it as follows:

> • Bouncer at Pete's Tavern, a local bar

Other places to look for series of items that should be parallel include outlines, procedures, and sets of subheadings within the same document.

Transitions within and between paragraphs

You can choose from three techniques to achieve smooth transitions within and between paragraphs.

(1) Use transitional expressions. These are words such as *again, furthermore, in addition, meanwhile, however, also, although, for example, specifically, in particular, as a result, in other words, certainly, accordingly, because, therefore.* Such words serve as bridges between ideas.

(2) Repeat key words, phrases, and concepts. Begin a sentence or paragraph by referring to something that was mentioned in the previous sentence or paragraph. The following paragraphs provide examples.

Original paragraph

> Breast cancer is a leading cause of death for women over 50. All women should learn to do monthly self-examinations. Doctors can easily teach women how to do these examinations, and public health organizations publish pamphlets to teach women how to do them. Many women think breast cancer will never happen to them, so they don't self-examinations.

Revised paragraph

> Breast cancer is a leading cause of death for women over 50. *Because breast cancer is so common and so deadly,* all women should learn to do monthly self-examinations. *Women* can easily *learn* to do these *examinations* from doctors or from pamphlets published by public health organizations. However, *even though self-examinations are easy to learn,* many women don't do them because they think breast cancer will never happen to them.

The italicized words in the revised paragraph indicate places where key information from one sentence is repeated in the sentence that follows. (Also note that the transitional phrase "however" is inserted at the beginning of the last sentence).

(3) Use forecasting statements to tell the reader where you are going next. The following list provides examples of forecasting statements.

> "The next step is to further examine Johnson's points. (*The writer would then proceed to examine Johnson's points.*)
>
> Of course we can also explore other avenues. (*The writer would then proceed to explore these avenues.*)
>
> There are at least two reasons why affirmative action as we know it will never improve racial relations in the United States." (*The writer would then proceed to list and explain these reasons.*)

Use these methods as revision guidelines, but remember that they will not solve every organizational problem that occurs in real-life writing situations. If you encounter a situation where you think you need a better transition, but none of these methods seems appropriate, it may be that you need to delete or move some information. Sometimes writers include information that is irrelevant to their topic, and sometimes information in one paragraph would tie in more easily with a different paragraph. In these situations, better transitions are not possible without doing some major renovation first.

Grammar topic 7: Mechanics

The mechanical aspects of writing a document include abbreviation, hyphenation, capitalization, use of numbers, and spelling. (Keep in mind that not all of these rules are hard and fast; some may depend on style guides used in your field or your company.)

Abbreviations

The following should *always* be abbreviated:

- Titles such as Ms., Mr., Dr., Jr. when they are used before or after a proper name.
- Time designations that are specific (400 BCE, 5:15 a.m.).

The following should *never* be abbreviated:

- Military, religious, or political titles (Reverend, President).
- Time designations that are used without actual times (Sarah arrived early in the morning—not early in the a.m.).

Avoid abbreviations whose meanings might not be clear to all readers. Units of measurement can be abbreviated if they appear frequently in your report. However, a unit of measurement should be spelled out the first time it is used. Avoid abbreviations in visual aids unless saving space is absolutely necessary.

Hyphenation

Hyphens divide words at line breaks and join two or more words used as a single adjective if they precede the noun (but not if they follow it):

> Com-puter (at a line break)
> The rough-hewn wood
> The all-too-human error
> The wood was rough hewn.
> The error was all too human.

Other commonly hyphenated words include the following:

- Most words that begin with the prefix self- (self-reliance, self-discipline— see your dictionary for exceptions).
- Combinations that might be ambiguous (re-creation versus recreation)
- Words that begin with *ex* only if *ex* means "past" (ex-faculty member, *but* excommunicate).
- All fractions, along with ratios that are used as adjectives and that precede the noun, and compound numbers from twenty-one through ninety-nine (a two-thirds majority, thirty-eight windows). Do not hyphenate fractions used as nouns (one third of the applicants).

Capitalization

Capitalize the first words of all sentences as well as titles of people, books, and chapters; languages; days of the week; the months; holidays; names of organizations or groups; races and nationalities; historical events; important documents; and names of structures or vehicles. In titles of books, films, and the like, capitalize the first words and all those following except articles or prepositions.

Do not capitalize the seasons (spring, winter) or general groups (the younger generation, the leisure class).

Capitalize adjectives derived from proper nouns (Chaucerian English).

Capitalize words such as street, road, corporation, and college only when they accompany a proper noun (Bob Jones University, High Street, The Rand Corporation).

Capitalize north, south, east, and west when they denote specific locations (the South, the Northwest), but not when they are simply directions (turn east at the light).

Use of numbers

Numbers expressed in one or two words can be written out or written as numerals. Use numerals to express larger numbers, decimals, fractions, precise technical figures, or any other exact measurements.

543	2,800,357
3.25	15 pounds of pressure
50 kilowatts	4,000 rpm

Use numerals for dates, census figures, addresses, page numbers, exact units of measurement, percentages, times with a.m. or p.m. designations, and monetary and mileage figures.

page 14	1:15 p.m.
18.4 pounds	9 feet
12 gallons	$15

Do not begin a sentence with a numeral. If your figure needs more than two words, revise your word order.

Six hundred students applied for the 102 available jobs.

The 102 available jobs brought 650 applicants.

Do not use numerals to express approximate figures, time not designated as a.m. or p.m., or streets named by numbers less than 100.

About seven hundred fifty

Four fifteen

108 East Forty-Second Street

In contracts and other documents in which precision is vital, a number can be stated both in numerals and in words:

> The tenant agrees to pay a rental fee of three hundred seventy-five dollars ($375.00) monthly.

Spelling

Always use the spell-check function in your word-processing software. However, don't rely on it exclusively. Take the time to use your dictionary for all writing assignments. And, if you are a poor speller, ask someone else to proofread every document before you present it to your primary audience.

Appendix B

Documenting Sources

Quoting the work of others

How you should document

MLA documentation style

APA documentation style

Quoting the work of others

You must place quotation marks around all exact wording you borrow, whether the words were written, spoken (as in an interview or presentation), or appeared in electronic form. Even a single borrowed sentence or phrase, or a single word used in a special way, needs quotation marks, with the exact source properly cited. These sources include people with whom you collaborate.

If your notes don't identify quoted material accurately, you might forget to credit the source. Even when this omission is unintentional, you face the charge of *plagiarism* (misrepresenting someone else's words or ideas as your own). Possible consequences of plagiarism include expulsion from school, loss of a job, and lawsuits.

Research writing is a process of independent thinking in which you work with the ideas of others in order to reach your own conclusions; unless the author's exact wording is essential, try to paraphrase instead of quoting borrowed material.

Paraphrasing the work of others

Paraphrasing means more than changing or shuffling a few words; it means restating the original idea in your own words—sometimes in a clearer, more direct, and emphatic way—and giving full credit to the source.

To borrow or adapt someone else's ideas or reasoning without properly documenting the source is plagiarism. To offer as a paraphrase an original passage that is only slightly altered—even when you document the source—is also plagiarism. Equally unethical is offering a paraphrase, although documented, that distorts the original meaning.

What you should document

Document any insight, assertion, fact, finding, interpretation, judgment, or other "appropriated material that readers might otherwise mistake for your own" (Gibaldi & Achtert, 1988, p. 155)—whether the material appears in published form or not. Specifically, you must document

- Any source from which you use exact wording
- Any source from which you adapt material in your own words
- Any visual illustration: charts, graphs, drawings, or the like (See Chapter 8 for documenting visuals.)

In some instances, you might have reason to preserve the anonymity of unpublished sources ("A number of employees expressed frustration with . . . "):

for example, to allow people to respond candidly without fear of reprisal (as with employee criticism of the company), or to protect their privacy (as with certain material from email inquiries or electronic newsgroups). You must still document the fact that you are not the originator of this material by providing a general acknowledgment in the text ("Interviews with Polex employees, May 1999") but not in your list of references.

You don't need to document anything considered *common knowledge:* material that appears repeatedly in general sources. In medicine, for instance, it has become common knowledge that foods containing animal fat contribute to higher blood cholesterol levels. So in a report on fatty diets and heart disease, you would probably not need to document that well-known fact. But you would document information about how the fat/cholesterol connection was discovered, what subsequent studies have found (say, the role of saturated versus unsaturated fats), and any information for which some other person or group could claim specific credit. If the borrowed material can be found in only one specific source, not in multiple sources, document it. When in doubt, document the source.

How you should document

Cite borrowed material twice: at the exact place you use that material, and at the end of your document. Documentation practices vary widely, but all systems work almost identically: A brief reference in the text names the source and refers readers to the complete citation, which allows readers to retrieve the source.

Many disciplines, institutions, and organizations publish their own style guides or documentation manuals. Here are a few:

Geographical Research and Writing

Style Manual for Engineering Authors and Editors

IBM Style Manual

NASA Publications Manual

Two common documentation styles are MLA (used mostly in the humanities) and APA (used mostly in the social sciences). Consult the most recent edition of *MLA Style Manual and Guide to Scholarly Publishing* or *Publication Manual of the American Psychological Association* for guidance on documenting sources according to these systems. For information on a documentation style designed specifically for electronic and Internet sources, see *The Columbia Guide to Online Style* (1998) by Janice R. Walker and Todd Taylor.

MLA documentation style

Traditional MLA documentation of sources used superscript numbers (like this:[1]) in the text, followed by full references at the bottom of the page (footnotes) or at the end of the document (endnotes) and, finally, by a bibliography. But a more current form of documentation appears in the *MLA Style Manual and Guide to Scholarly Publishing*, 2nd ed., New York: Modern Language Association, 1998. Footnotes or endnotes are now used only to comment on material in the text or on sources or to suggest additional sources.

In current MLA style, in-text parenthetical references briefly identify the source(s). Full documentation then appears in a Works Cited section at the end of the document.

A parenthetical reference usually includes the author's surname and the exact page number(s) of the borrowed material:

> One notable study indicates an elevated risk of leukemia for children exposed to certain types of electromagnetic fields (Bowman et al. 59).

Readers seeking the complete citation for Bowman can refer easily to the Works Cited section, listed alphabetically by author:

> Bowman, J. D., et al. "Hypothesis: The Risk of Childhood Leukemia Is Related to Combinations of Power-Frequency and Static Magnetic Fields." *Bioelectromagnetics* 16.1 (1995): 48–59.

This complete citation includes page numbers for the entire article.

MLA parenthetical references

For clear and informative parenthetical references, observe these rules:

- If your discussion names the author, do not repeat the name in your parenthetical reference; simply give the page number(s):

> Bowman et al. explain how their recent study indicates an elevated risk of leukemia for children exposed to certain types of electromagnetic fields (59).

- If you cite two or more works in a single parenthetical reference, separate the citations with semicolons:

> (Jones 32; Leduc 41; Gomez 293–94)

- If you cite two or more authors with same surnames, include the first initial in your parenthetical reference to each author:

(R. Jones 32)

(S. Jones 14–15)

- If you cite two or more works by the same author, include the first significant word from each work's title, or a shortened version:

(Lamont, *Biophysics* 100–01)

(Lamont, *Diagnostic Tests* 81)

- If the work is by an institutional or corporate author or if it is unsigned (that is, the author is unknown), use only the first few words of the institutional name or the work's title in your parenthetical reference:

(American Medical Assn. 2)

("Distribution Systems" 18)

To avoid distracting the reader, keep each parenthetical reference brief. The easiest way to keep parenthetical references brief is to name the source in your discussion and place only the page number(s) in parentheses.

For a paraphrase, place the parenthetical reference *before* the closing punctuation mark. For a quotation that runs into the text, place the reference *between* the final quotation mark and the closing punctuation mark. For a quotation set off (indented) from the text, place the reference two spaces *after* the closing punctuation mark.

MLA works cited entries

The Works Cited list includes each source that you have paraphrased or quoted in your document. In preparing the list, type the first line of each entry flush with the left margin. Indent the second and subsequent lines five spaces. Double-space within and between each entry. Use one character space after any period, comma, or colon.

Following are examples of complete citations as they would appear in the Works Cited section of your document. Shown after each citation is the corresponding parenthetical reference as it would appear in the text.

MLA WORKS CITED ENTRIES FOR BOOKS. Any citation for a book should contain the following information: author, title, editor or translator, edition, volume number, and facts about publication (city, publisher, date).

INDEX TO SAMPLE MLA WORKS CITED ENTRIES

Books

1. Book, single author
2. Book, two or three authors
3. Book, four or more authors
4. Book, anonymous author
5. Multiple books, same author
6. Book, one or more editors
7. Book, indirect source
8. Anthology selection or book chapter

Periodicals

9. Article, magazine
10. Article, journal with new pagination each issue
11. Article, journal with continuous pagination
12. Article, newspaper

Other Sources

13. Encyclopedia, dictionary, alphabetic reference
14. Report
15. Conference presentation

16. Interview, personally conducted
17. Interview, published
18. Letter, unpublished
19. Questionnaire
20. Brochure or pamphlet
21. Lecture
22. Government document
23. Document with corporate authorship
24. Map or other visual
25. Unpublished dissertation, report or miscellaneous items

Electronic Sources

26. Online database
27. Computer software
28. CD-ROM
29. Listserv
30. Usenet
31. Email
32. Web site
33. Article in online periodical
34. Real-time communication

1. Book, Single Author—MLA

Kerzin-Fontana, Jane B. *Technology Management: A Handbook.* 3rd ed. Delmar, NY: American Management Assn., 1999.

Parenthetical reference: (Kerzin-Fontana 3–4)

Identify the state of publication by U.S. Postal Service abbreviations. If the city of publication is well known (Boston, Chicago), omit the state abbreviations. If several cities are listed on the title page, give only the first. For Canada, include the province abbreviation after the city. For all other countries include an abbreviation of the country name.

2. Book, Two or Three Authors—MLA

Aronson, Linda, Roger Katz, and Candide Moustafa. *Toxic Waste Disposal Methods.* New Haven: Yale UP, 2000.

Parenthetical reference: (Aronson, Katz, and Moustafa 121–23)

Shorten publisher's names, as in "Simon" for Simon & Schuster, "GPO" for Government Printing Office, or "Yale UP" for Yale University Press. For page numbers with more than two digits, give only the final digits for the second number.

3. Book, Four or More Authors—MLA

Santos, Ruth J., et al. *Environmental Crises in Developing Countries.* New York: Harper, 1998.

Parenthetical reference: (Santos et al. 9)

"Et al." is the abbreviated form of the Latin "et alia," meaning "and others."

4. Book, Anonymous Author—MLA

Structured Programming. Boston: Meredith, 1999.

Parenthetical reference: (*Structured* 67)

5. Multiple Books, Same Author—MLA

Chang, John W. *Biophysics.* Boston: Little, 1999.

———. *Diagnostic Techniques.* New York: Radon, 1994.

Parenthetical references: (Chang, *Biophysics* 123–26), (Chang, *Diagnostic* 87)

When citing more than one work by the same author, do not repeat the author's name; type three hyphens followed by a period. List the works alphabetically by title.

6. Book, One or More Editors—MLA

Morris, A. J., and Louise B. Pardin-Walker, eds. *Handbook of New Information Technology.* New York: Harper, 1996.

Parenthetical reference: (Morris and Pardin-Walker 34)

For more than three editors, name only the first, followed by "et al."

7. Book, Indirect Source—MLA

Kline, Thomas. *Automated Systems.* Boston: Rhodes, 1992.

Stubbs, John. *White-Collar Productivity.* Miami: Harris, 1999.

Parenthetical reference: (qtd. in Stubbs 116)

When your source (as in Stubbs, above) has quoted or cited another source, list each source in its appropriate alphabetical place on your Works Cited page. Use the name of the original source (here, Kline) in your text and precede your parenthetical reference with "qtd. in," or "cited in" for a paraphrase.

8. Anthology Selection or Book Chapter—MLA

> Bowman, Joel P. "Electronic Conferencing." *Communication and Technology: Today and Tomorrow.* Ed. Al Williams. Denton, TX: Assn. for Business Communication, 1994. 123–42.
>
> *Parenthetical reference:* (Bowman 129)

The page numbers in the complete citation are for the selection cited from the anthology.

MLA WORKS CITED ENTRIES FOR PERIODICALS. A citation for an article should give this information (as available): author, article title, periodical title, volume or number (or both), date (day, month, year), and page numbers for the entire article—not just the pages cited. List the information in this order, as in the following examples.

9. Article, Magazine—MLA

> DesMarteau, Kathleen. "Study Links Sewing Machine Use to Alzheimer's Disease." *Bobbin* Oct. 1994: 36–38.
>
> *Parenthetical reference:* (DesMarteau 36)

No punctuation separates the magazine title and date. Nor is the abbreviation "p." or "pp." used to designate page numbers.

If no author is given, list all other information:

> Distribution Systems for the New Decade. *Power Technology Magazine* 18 Oct. 2000: 18+.
>
> Parenthetical reference: ("Distribution Systems" 18)

This article begins on page 18 and continues on page 21. When an article does not appear on consecutive pages, give only the number of the first page, followed immediately by a plus sign. A three-letter abbreviation denotes any month spelled with five or more letters.

10. Article, Journal with New Pagination Each Issue—MLA

> Thackman-White, Joan R. "Computer-Assisted Research." *American Librarian* 51.1 (1999): 3–9.
>
> *Parenthetical reference:* (Thackman-White 4–5)

Because each issue for a given year will have page numbers beginning with "1," readers need the issue number. The "51" denotes the volume number; "1" denotes the issue number. Omit "The" or "A" or any other introductory article from a journal or magazine title.

11. Article, Journal with Continuous Pagination—MLA

Barnstead, Marion H. "The Writing Crisis." *Journal of Writing Theory* 12 (1998): 415–33.

Parenthetical reference: (Barnstead 415–16)

When page numbers continue from one issue to the next for the full year, readers won't need the issue number, because no other issue in that year repeats these same page numbers. (Include the issue number, however, if you think it will help readers retrieve the article.) The "12" denotes the volume number.

12. Article, Newspaper—MLA

Baranski, Vida H. "Errors in Technology Assessment." *Boston Times* 15 Jan. 1999, evening ed., sec. 2: 3.

Parenthetical reference: (Baranski 3)

When a daily newspaper has more than one edition, cite the edition after the date. Omit any introductory article in the newspaper's name (not *The Boston Times*). If no author is given, list all other information. If the newspaper's name does not include the city of publication, insert it, using brackets: *Sippican Sentinel* [Marion, MA].

MLA WORKS CITED ENTRIES FOR OTHER KINDS OF MATERIALS. Miscellaneous sources range from unsigned encyclopedia entries to conference presentations to government publications. A full citation should give this information (as available): author, title, city, publisher, date, and page numbers.

13. Encyclopedia, Dictionary, Other Alphabetical Reference—MLA

"Communication." *The Business Reference Book,* 1998 ed.

Parenthetical reference: ("Communication")

Begin a signed entry with the author's name. For any work arranged alphabetically, omit page numbers in the citation and the parenthetical reference. For a well-known reference book, include only an edition (if stated) and a date. For other reference books, give the full publication information.

14. Report—MLA

Electrical Power Research Institute (EPRI). *Epidemiologic Studies of Electric Utility Employees.* (Report No. RP2964.5). Palo Alto, CA: EPRI, Nov. 1994.

Parenthetical reference: (Electrical Power Research Institute [EPRI] 27)

If no author is given, begin with the organization that sponsored the report.

For any report or other document with group authorship, as above, include the group's abbreviated name in your first parenthetical reference, and then use only that abbreviation in any subsequent reference.

15. Conference Presentation—MLA

Smith, Abelard A. "Radon Concentrations in Molded Concrete." *First British Symposium in Environmental Engineering. London, 11–13 Oct. 1998.* Ed. Anne Hodkins. London: Harrison, 1999. 106–21.

Parenthetical reference: (Smith 109)

The above example shows a presentation that has been included in the published proceedings of a conference. For an unpublished presentation, include the presenter's name, the title of the presentation, and the conference title, location, and date, but do not underline or italicize the conference information.

16. Interview, Personally Conducted—MLA

Nasser, Gamel. Chief Engineer for Northern Electric. Personal interview. Rangeley, ME. 2 Apr. 1999.

Parenthetical reference: (Nasser)

17. Interview, Published—MLA

Lescault, James. "The Future of Graphics." *Executive Views of Automation.* Ed. Karen Prell. Miami: Haber, 2000. 216–31.

Parenthetical reference: (Lescault 218)

The interviewee's name is placed in the entry's author slot.

18. Letter, Unpublished—MLA

Rogers, Leonard. Letter to the author. 15 May 1998.

Parenthetical reference: (Rogers)

19. Questionnaire—MLA

Taylor, Lynne. Questionnaire sent to 612 Massachusetts business executives. 14 Feb. 2000.

Parenthetical reference: (Taylor)

20. Brochure or Pamphlet—MLA

Investment Strategies for the 21st Century. San Francisco: Blount Economics Assn., 1999.

Parenthetical reference: (*Investment*)

If the work is signed, begin with its author.

21. Lecture—MLA

Dumont, R. A. "Managing Natural Gas." Lecture. University of Massachusetts at Dartmouth, 15 Jan. 1998.

Parenthetical reference: (Dumont)

If the lecture title is not known, write Address, Lecture, or Reading but do not use quotation marks. Include the sponsor and the location if available.

22. Government Document—MLA

Virginia. Highway Dept. *Standards for Bridge Maintenance.* Richmond: Virginia Highway Dept., 1997.

Parenthetical reference: (Virginia Highway Dept. 49)

If the author is unknown (as above), list the information in this order: name of the government, name of the issuing agency, document title, place, publisher, and date.

For any congressional document, identify the house of Congress (Senate or House of Representatives) before the title, and the number and session of Congress after the title:

United States Cong. House. Armed Services Committee. *Funding for the Military Academies.* 105th Congress, 2nd sess. Washington: GPO, 1998.

Parenthetical reference: (U.S. Cong. 41)

"GPO" is the abbreviation for the U.S. Government Printing Office.

For an entry from the *Congressional Record,* give only date and pages:

Cong. Rec. 10 Mar. 1999: 2178–92.

Parenthetical reference: (*Cong. Rec.* 2184)

23. Document with Corporate Authorship—MLA

Hermitage Foundation. *Global Warming Scenarios for the Year 2030.* Washington: Natl. Res. Council, 2000.

Parenthetical reference: (Hermitage Foun. 123)

24. Map or Other Visual—MLA

Deaths Caused by Breast Cancer, by County. Map. *Scientific American* Oct. 1995: 32D.

Parenthetical reference: (*Deaths Caused*)

If the creator of the visual is listed, give that name first. Identify the type of visual (Map, Graph, Table, Diagram) immediately following the title.

25. Unpublished Dissertation, Report, or Miscellaneous Items—MLA

> Author (if known), "Title." Sponsoring organization or publisher, date, page numbers.

For any work that has group authorship (corporation, committee, task force),
cite the name of the group or agency in place of the author's name.

MLA Works cited entries for electronic sources. Citation for an electronic source
with a printed equivalent should begin with that publication information (see
relevant sections above). But whether or not a printed equivalent exists, any
citation should enable readers to retrieve the material electronically.

The Modern Language Association recommends these general conventions:

Publication Dates: For sources taken from the Internet, include the date the
source was posted to the Internet or last updated or revised as well as the date
you accessed the source.

Uniform Resource Locators: Include a full and accurate URL (electronic
address) for any source taken from the Internet (with access mode identifier—
http, ftp, gopher, or *telnet*). Enclose URLs in angle brackets (< >). When a
URL continues from one line to the next, break it only after a slash. Do not
add a hyphen.

Page Numbering: Include page or paragraph numbers when given by the
source.

26. Online Database—MLA

> Sahl, J. D. "Power Lines, Viruses, and Childhood Leukemia." *Cancer Causes Control* 6.1
> (Jan. 1995): 83. *MEDLINE.* Online. DIALOG. 7 Nov. 1995.
>
> *Parenthetical reference:* (Sahl 83)

For entries with a printed equivalent, begin with publication information,
then the database title (underlined or italicized), the "Online" designation to
indicate the medium, and the service provider (or URL or email address) and
the date of access. The access date is important because frequent updatings of
databases can produce different versions of the material.

For entries with no printed equivalent, give the title and date of the work in
quotation marks, followed by the electronic source information:

> Argent, Roger R. "An Analysis of International Exchange Rates for 1999." *Accu-Data.*
> Online. Dow Jones News Retrieval. 10 Jan. 2000.
>
> *Parenthetical reference:* (Argent 4)

If the author is not known, begin with the work's title.

27. Computer Software—MLA

Virtual Collaboration. Diskette. New York: Harper, 1994.

Parenthetical reference: (*Virtual*)

Begin with the author's name, if known.

28. CD-ROM—MLA

Cavanaugh, Herbert A. "EMF Study: Good News and Bad News."

Electrical World Feb. 1995: 8. *ABI/INFORM.* CD-ROM. Proquest. Sept. 1995.

Parenthetical reference: (Cavanaugh 8)

If the material is also available in print, begin with the information about the printed source, followed by the electronic source information: name of the database (underlined), CD-ROM designation, vendor name, and electronic publication date. If the material has no printed equivalent, list its author (if known) and title (in quotation marks), followed by the electronic source information.

If you are citing an abstract of the complete work, insert "Abstract," followed by a period, immediately after the work's page number(s)—"8" in the previous entry.

For CD-ROM reference works and other material not routinely updated, give the title of the work, followed by the CD-ROM designation, place, electronic publisher, and date:

Time Almanac. CD-ROM. Washington: Compact, 1994.

Parenthetical reference: (*Time Almanac* 74)

Begin with the author's name, if known.

29. Listserv—MLA

Korsten, A. "Major Update of the WWWVL Migration and Ethnic Relations." 7 Apr. 1998. Online posting. ERCOMER News. 8 Apr. 1998

<www.ercomer.org/archive/ercomer-news/0002.html>.

Parenthetical reference: (Korsten)

Begin with the author's name (if known), followed by the title of the work (in quotation marks), publication date, the Online posting designation, name of discussion group, date of access, and the URL. The parenthetical reference includes no page number because none is given in an online posting.

30. Usenet—MLA

Dorsey, Michael. "Enviromentalism or Racism." 25 Mar. 1998. Online posting. 1 Apr. 1998 <news:alt.org.sierra-club>.

Parenthetical reference: (Dorsey)

31. Email—MLA

Wallin, John Luther. "Frog Reveries." email to the author. 12 Oct. 1999.

Parenthetical reference: (Wallin)

Cite personal email as you would printed correspondence. If the document has a subject line or title, enclose it in quotation marks.

For publicly posted email (say, a newsgroup or discussion list) include the address and date of access.

32. Web Site—MLA

Dumont, R. A. "An Online Course in Technical Writing." 10 Dec. 1999. UMASS Dartmouth Online. 6 Jan. 2000

<www.umassd.edu/englishdepartment.html>.

Parenthetical reference: (Dumont 7–9)

Begin with the author's name (if known), followed by title of the work (in quotation marks), posting date, name of Web site, date of access, and Web address (in angle brackets). Note that a Web address that continues from one line to the next is broken only after the slash(es). No hyphen is added.

33. Article in an Online Periodical—MLA

Jones, Francine L. "The Effects of NAFTA on Labor Union Membership." *Cambridge Business Review* 2.3 (1999): 47–64. 4 Apr. 2000 <www.mun.ca/cambrbusrev/1999vol2/jones2.html>.

Parenthetical reference: (Jones 44–45)

Information about the printed version is followed by the date of access to the Web site and the electronic address.

34. Real-Time Communication—MLA

Synchronous communication occurs in a "real time" forum and includes MUDs (multiuser dungeons), MOOs (MUD object-oriented software), IRC (Internet relay chat), and FTPs (file transfer protocols). The message typed in by the sender appears instantly on the screen of the recipient, as in a personal interview.

Mendez, Michael R. Online debate. "Solar power versus Fossil Fuel Power." 3 Apr. 1998. CollegeTownMoo. 3 Apr. 1998 <telnet://next.cs.bvc.edu.777>.

Parenthetical reference: (Mendez)

Begin with the name of the communicator(s) and indicate the type of communication (personal interview, online debate, and so on), topic title, posting date, name of forum, access date, and electronic address.

MLA works cited page

Place your Works Cited section on a separate page at the end of the document. Arrange entries alphabetically by author's surname. When the author is unknown, list the title alphabetically according to its first word (excluding introductory articles). For a title that begins with a digit ("5," "6," etc.), alphabetize the entry as if the digit were spelled out.

APA documentation style

One popular alternative to MLA style appears in the *Publication Manual of the American Psychological Association,* 4th ed., Washington: American Psychological Association, 1994. APA style is useful when writers wish to emphasize the publication dates of their references. A parenthetical reference in the text briefly identifies the source, date, and page number(s):

In a recent study, mice continuously exposed to an electromagnetic field tended to die earlier than mice in the control group (de Jager & de Bruyn, 1994, p. 224).

The full citation then appears in the alphabetical listing of "References," at the report's end:

de Jager, L., & de Bruyn, L. (1994). Long-term effects of a 50 Hz electric field on the life expectancy of mice. *Review of Environmental Health, 10* (3–4), 221–224.

Because it emphasizes the date, APA style (or some similar author-date style) is preferred in the sciences and social sciences, where information quickly becomes outdated.

APA parenthetical references

APA's parenthetical references differ from MLA's as follows: The APA citation includes the publication date; a comma separates each item in the reference; and "p." or "pp." precedes the page number (which is optional in the APA system).

When a subsequent reference to a work follows closely after the initial reference, the date need not be included. Here are specific guidelines:

- If your discussion names the author, do not repeat the name in your parenthetical reference; simply give the date and page number:

    Researchers de Jager and de Bruyn explain that experimental mice exposed to an electromagnetic field tended to die earlier than mice in the control group (1994, p. 224).

    When two authors of a work are named in the text, their names are connected by "and," but in a parenthetical reference, their names are connected by an ampersand, "&."

- If you cite two or more works in a single reference, list the authors in alphabetical order and separate the citations with semicolons:

    (Jones, 1994; Gomez, 1992; Leduc, 1996)

- If you cite a work with three to five authors, try to name them in your text, to avoid an excessively long parenthetical reference.

    Franks, Oblesky, Ryan, Jablar, and Perkins (1993) studied the role of electromagnetic fields in tumor formation.

    In any subsequent references to this work, name only the first author, followed by "et al." (Latin abbreviation for "and others").

- If you cite two or more works by the same author published in the same year, assign a different letter to each work:

    (Lamont, 1990a, p. 135)
    (Lamont, 1990b, pp. 67–68)

Other examples of parenthetical references appear with their corresponding entries in the following discussion of the list of references.

APA reference list entries

The APA reference list includes each source that you have cited in your document. In preparing the list of references, type the first line of each entry flush with the left margin. Indent the second and subsequent lines five spaces. Skip one character space after any period, comma, or colon. Double-space within and between each entry.

Following are examples of complete citations as they would appear in the References section of your document. Shown immediately below each entry is its corresponding parenthetical reference as it would appear in the text. Note the capitalization, abbreviation, spacing, and punctuation in the sample entries.

INDEX TO SAMPLE ENTRIES FOR APA REFERENCES

Books

1. Book, single author
2. Book, two to five authors
3. Book, six or more authors
4. Book, anonymous author
5. Multiple books, same author
6. Book, one to five editors
7. Book, indirect source
8. Anthology selection or book chapter

Periodicals

9. Article, magazine
10. Article, journal with new pagination each issue
11. Article, journal with continuous pagination
12. Article, newspaper

Other Sources

13. Encyclopedia, dictionary, alphabetical reference

14. Report
15. Conference presentation
16. Interview, personally conducted
17. Interview, published
18. Personal correspondence
19. Brochure or pamphlet
20. Lecture
21. Government document
22. Miscellaneous items

Electronic Sources

23. Online database abstract
24. Online database article
25. Computer software or software manual
26. CD-ROM abstract
27. CD-ROM reference work
28. Electronic bulletin board, discussion list, email
29. Web site

APA ENTRIES FOR BOOKS. Any citation for a book should contain all applicable information in the following order: author, date, title, editor or translator, edition, volume number, and facts about publication (city and publisher).

1. Book, Single Author—APA

> Kerzin-Fontana, J. B. (1999). *Technology management: A handbook* (3rd ed.). Delmar, NY: American Management Association.

> *Parenthetical reference:* (Kerzin-Fontana, 1999, pp. 3–4)

Use only initials for an author's first and middle name. Capitalize only the first words of a book's title and subtitle and any proper names. Identify a later edition in parentheses between the title and the period.

2. Book, Two to Five Authors—APA

> Aronson, L., Katz, R., & Moustafa, C. (2000). *Toxic waste disposal methods.* New Haven: Yale University Press.
>
> *Parenthetical reference:* (Aronson, Katz, & Moustafa, 2000)

Use an ampersand (&) before the name of the final author listed in an entry. As an alternative parenthetical reference, name the authors in your text and include date (and page numbers, if appropriate) in parentheses.

3. Book, Six or More Authors—APA

> Fogle, S. T., et al. (1998). *Hyperspace technology.* Boston: Little, Brown.
>
> *Parenthetical reference:* (Fogle et al., 1998, p. 34)

"Et al." is the Latin abbreviation for "et alia," meaning "and others."

4. Book, Anonymous Author—APA

> *Structured programming.* (1995). Boston: Meredith Press.
>
> *Parenthetical reference:* (*Structured Programming,* 1995, p. 67)

In your list of references, place an anonymous work alphabetically by the first key word (not *The, A,* or *An*) in its title. In your parenthetical reference, capitalize all key words in a book, article, or journal title.

5. Multiple Books, Same Author—APA

> Chang, J. W. (1997a). *Biophysics.* Boston: Little, Brown.
>
> Chang, J. W. (1997b). *MindQuest.* Chicago: John Pressler.
>
> *Parenthetical references:* (Chang, 1997a)
> (Chang, 1997b)

Two or more works by the same author not published in the same year are distinguished by their respective dates alone, without the added letter.

6. Book, One to Five Editors—APA

> Morris, A. J., & Pardin-Walker, L. B. (Eds.). (1999). *Handbook of new information technology.* New York: HarperCollins.
>
> *Parenthetical reference:* (Morris & Pardin-Walker, 1999, p. 79)

For more than five editors, name only the first, followed by "et al."

7. Book, Indirect Source—APA

> Stubbs, J. (1998). *White-collar productivity.* Miami: Harris.
>
> *Parenthetical reference:* (cited in Stubbs, 1998, p. 47)

When your source (as in Stubbs, above) has cited another source, list only this second source in the References section, but name the original source in the text: "Kline's study (cited in Stubbs, 1998, p. 47) supports this conclusion."

8. Anthology Selection or Book Chapter—APA

> Bowman, J. (1994). Electronic conferencing. In A. Williams (Ed.), *Communication and technology: Today and tomorrow* (pp. 123–142). Denton, TX: Association for Business Communication.

> *Parenthetical reference:* (Bowman, 1994, p. 126)

The page numbers in the complete reference are for the selection cited from the anthology.

APA ENTRIES FOR PERIODICALS. A citation for an article should give this information (as available), in order: author, publication date, article title (without quotation marks), volume or number (or both), and page numbers for the entire article—not just the page(s) cited.

9. Article, Magazine—APA

> DesMarteau, K. (1994, October). Study links sewing machine use to Alzheimer's disease. *Bobbin, 36,* 36–38.

> *Parenthetical reference:* (DesMarteau, 1994, p. 36)

If no author is given, provide all other information. Capitalize the first word in an article's title and subtitle, and any proper nouns. Capitalize all key words in a periodical title. Underline or italicize the periodical title, volume number, and commas (as above).

10. Article, Journal with New Pagination for Each Issue—APA

> Thackman-White, J. R. (1999). Computer-assisted research. *American Library Journal, 51*(1), 3–9.

> *Parenthetical reference:* (Thackman-White, 1999, pp. 4–5)

Because each issue for a given year has page numbers that begin at "1," readers need the issue number (in this instance, "1"). The "51" denotes the volume number, which is underlined or italicized.

11. Article, Journal with Continuous Pagination—APA

> Barnstead, M. H. (1999). "The Writing Crisis." *Journal of Writing Theory 12,* 415–433.

> *Parenthetical reference:* (Barnstead, 1999, pp. 415–416)

The "12" denotes the volume number. When page numbers continue from issue to issue for the full year, readers won't need the issue number, because no other issue in that year repeats these same page numbers. (You can still include the issue number if you think it will help readers retrieve the article more easily.)

12. Article, Newspaper—APA

> Baranski, V. H. (1999, January 15). Errors in technology assessment. *The Boston Times,* p. B3.

> *Parenthetical reference:* (Baranski, 1999, p. B3)

In addition to the year of publication, include the month and day. If the newspaper's name begins with "The," include it in your citation. Include "p." or "pp." before page numbers. For an article on nonconsecutive pages, list each page, separated by a comma.

APA ENTRIES FOR OTHER SOURCES. Miscellaneous sources range from unsigned encyclopedia entries to conference presentations to government documents. A full citation should give this information (as available): author, publication date, work title (and report or series number), page numbers (if applicable), city, and publisher.

13. Encyclopedia, Dictionary, Alphabetic Reference—APA

> Communication. (1998). In *The business reference book.* Boston: Business Resources Press.

> *Parenthetical reference:* ("Communication," 1998)

For an entry that is signed, begin with the author's name and publication date.

14. Report—APA

> Electrical Power Research Institute. (1994). *Epidemiologic studies of electric utility employees* (Report No. RP2964.5). Palo Alto, CA: Author.

> *Parenthetical reference:* (Electrical Power Research Institute [EPRI], 1994, p. 12)

If authors are named, list them first, followed by the publication date. When citing a group author, as above, include the group's abbreviated name in your first parenthetical reference, and use only that abbreviation in any subsequent reference. When the agency (or organization) and publisher are the same, list "Author" in the publisher's slot.

15. Conference Presentation—APA

> Smith, A. A. (1999). Radon concentrations in molded concrete. In A. Hodkins (Ed.), *First British Symposium on Environmental Engineering* (pp. 106–121). London: Harrison Press, 2000.

> *Parenthetical reference:* (Smith, 1999, p. 109)

In parentheses is the date of the presentation. The name of the symposium is a proper name, and so is capitalized. Following the publisher's name is the date of publication.

For an unpublished presentation, include the presenter's name, year and month, title of the presentation (underlined or italicized), and all available information about the conference or meeting: "Symposium held at. . . ." Do not underline or italicize this last information.

16. Interview, Personally Conducted—APA

Parenthetical reference: (G. Nasser, personal interview, April 2, 1999)

This material is considered a nonrecoverable source, and so is cited in the text only, as a parenthetical reference. If you name the respondent in text, do not repeat the name in the citation.

17. Interview, Published—APA

Jable, C. K. (1998). The future of graphics [Interview with James Lescault]. In K. Prell (Ed.), *Executive Views of Automation* (pp. 216–231). Miami: Haber Press, 1999.

Parenthetical reference: (Jable, 2000, pp. 218–223)

Begin with the name of the interviewer, followed by the interview date and title (if available), the designation (in brackets), and the publication information, including the date.

18. Personal Correspondence—APA

Parenthetical reference: (L. Rogers, personal correspondence, May 15, 1998)

This material is considered nonrecoverable data, and so is cited in the text only, as a parenthetical reference. If you name the correspondent in text, do not repeat the name in the citation.

19. Brochure or Pamphlet—APA

This material follows the citation format for a book entry (page 318). After the title of the work, include the designation "Brochure" in brackets.

20. Lecture—APA

Dumont, R. A. (1998, January 15). *Managing natural gas.* Lecture presented at the University of Massachusetts at Dartmouth.

Parenthetical reference: (Dumont, 1998)

If you name the lecturer in text, do not repeat the name in the citation.

21. Government Document—APA

Virginia Highway Department. (1997). *Standards for bridge maintenance.* Richmond: Author.

Parenthetical reference: (Virginia Highway Department, 1997, p. 49)

If the author is unknown, present the information in this order: name of the issuing agency, publication date, document title, place, and publisher. When the issuing agency is both author and publisher, list "Author" in the publisher's slot.

For any congressional document, identify the house of Congress (Senate or House of Representatives) before the date.

U.S. House Armed Services Committee. (1998). *Funding for the military academies.* Washington, DC: U.S. Government Printing Office.

Parenthetical reference: (U.S. House, 1998, p. 41)

22. Miscellaneous Items (unpublished manuscripts, dissertations, and so on)—APA

Author (if known). (Date of publication). *Title of work.* Sponsoring organization or publisher.

For any work that has group authorship (corporation, committee, and so on), cite the name of the group or agency in place of the author's name.

APA ENTRIES FOR ELECTRONIC SOURCES. APA documentation standards for electronic sources continue to be refined and defined. A sampling of currently preferred formats is presented below. Any citation for electronic media should allow readers to identify the original source (printed or electronic) and provide an electronic path for retrieving the material.

Begin with the publication information for the printed equivalent. Then, in brackets, name the electronic source ([Online], [CD-ROM], [Computer software]), the protocol[1] (Bitnet, Dialog, FTP, Telnet), and any other items that define a clear path (service provider, database title, access code, retrieval number, or site address).

23. Online Database Abstract—APA

Sahl, J. D. (1995). Power lines, viruses, and childhood leukemia [Online]. *Cancer Causes Control, 6* (1), 83. Abstract from: DIALOG File: *MEDLINE* Item: 93–04881

Parenthetical reference: (Sahl, 1995)

Note the absence of closing punctuation in items 23, 24, 26, and 29. Any punctuation added to the availability statement could interfere with retrieval.

24. Online Database Article—APA

Alley, R. A. (1999, January). Ergonomic influences on worker satisfaction [29 paragraphs]. *Industrial Psychology* [Online serial], *5* (11). Available FTP: Hostname: publisher.com Directory: pub/journals/ industrial.psychology/1999

Parenthetical reference: (Alley, 1995)

Give the length of the article [in paragraphs], after its title.

25. Computer Software or Software Manual—APA

Virtual collaboration [Computer software]. (1994). New York: HarperCollins.

Parenthetical reference: (Virtual, 1994)

---

[1]A *protocol* is a body of standards that ensures compatibility among the different products designed to work together on a particular network.

For citing a manual, replace the "Computer software" designation in brackets with "Software manual."

26. CD-ROM Abstract—APA

Cavanaugh, H. (1995). An EMF study: Good news and bad news [CD-ROM]. *Electrical World, 209*(2), 8. Abstract from: Proquest File: ABI/Inform: 978032

Parenthetical reference: (Cavanaugh, 1995)

The "8" in the entry above denotes the page number of this one-page article.

27. CD-ROM Reference Work—APA

Time almanac. (1994). Washington: Compact, 1994.

Parenthetical reference: (*Time almanac,* 1994)

If the work on CD-ROM has a printed equivalent, APA currently prefers that it be cited in its printed form. As more works appear in electronic form, this convention may be revised.

28. Electronic Bulletin Board, Discussion List, Email—APA

Parenthetical reference: Fred Flynn (personal communication, May 10, 1999) provided these statistics.

This material is considered personal communication in APA style. Instead of being included in the list of references, it is cited directly in the text. According to APA's current standards, material from discussion lists and electronic bulletin boards has limited research value because it does not undergo the kind of review and verification process used in scholarly publications.

29. Web Site—APA

Dumont, R. A. (1999). *An Online course in composition* [Online Web site]. Available WWW.www.umassd.edu/englishdepartment.html

If the Web address continues from one line to the next, divide it only after the slash(es).

APA references list

APA's References section is an alphabetical listing (by author) equivalent to MLA's Works Cited section. Like Works Cited, the reference list includes only those works actually cited. (A bibliography would usually include background works or works consulted as well.) Unlike MLA style, APA style calls for only "recoverable" sources to appear in the reference list. Therefore, personal interviews, email messages, and other unpublished materials are cited in the text only.

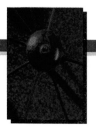

References

Blum, D. (1997). Investigative science journalism. In D. Blum and M. Knudson (Eds.), *Field guide for science writers* (pp. 86–93). New York: Oxford.

Brody, H. (1997, May/June). Clicking onto webzines. *Technology Review,* 40–47.

Cavazos, E. A., & Morin, G. (1994). *Cyberspace and the law: Your rights and duties in the on-line world.* Cambridge, MA: MIT Press.

Christians, C., Tackler, M., Rotzoll, K., & McKee, K. (eds.). (1978). *Media ethics: Cases in moral reasoning* (2nd ed.). New York: Longman.

Coe, M. (1996). *Human factors for technical communicators.* New York: Wiley.

Dragga, S. (1996). "Is this ethical?" A survey of opinion on principles and practices of document design. *Technical Communication, 43*(3), 255–265.

Dumas, J. S., & Redish, J. C. (1994). *A practical guide to usability testing* (2nd ed.). Norwood, NJ: Ablex.

Electronic Privacy Information Center (EPIC) (2000). www.epic.org.

Felker, D. B., Pickering, F., Charrow, V., Holland, V., & Redish, J. (1981). *Guidelines for document designers.* Washington: American Institutes for Research.

Fink, C. (1988). *Media ethics.* Boston: Allyn & Bacon.

Fortner, B., & Meyer, T. E. (1997). *Number by color: A guide to using color to understand technical data.* New York: Springer-Verlag.

Garfield, E. (1973). What scientific journals can tell us about scientific journals. *IEEE Transactions on Professional Communication, 16,* (4), 200–202.

Gibaldi, J., & Achtert, W. S. (1988). *MLA Handbook for Writers of Research Papers* (3rd ed.). New York: Modern Language Association.

Gouran, D. S., Hirokawa, R. Y., & Martz, A. E. (1986). A critical analysis of factors related to decisional processes involved in the *Challenger* disaster. *Central States Speech Journal, 37*(3), 119–135.

Gross, A., & Walzer, A. (1994). Positivists, postmodernists, Aristotelians, and the *Challenger* disaster. *College English, 56*(4), 420–433.

Hargis, G., Hernandez, A., Hughes, P., & Ramaker, J. (1997). *Developing quality technical information: A handbook for writers and editors.* Upper Saddle River, NJ: Prentice Hall.

Hartley, J. (1985). *Designing instructional text* (2nd ed.). London: Kogan Page.

Helyar, P. S., & Doudnikoff, G. M. (1994). Walking the labyrinth of multimedia law. *Technical Communication, 41*(4), 662–671.

Hoft, N. L. (1995). *International technical communication: How to export information about high technology.* New York: Wiley.

Horton, W. K. (1990). *Designing and writing online documentation: Help files to hypertext.* New York: Wiley.

Horton, William. (1991). *Illustrating computer documentation: The art of presenting information graphically on paper and online.* New York: Wiley.

Investor's Business Daily (1998, December 9). From Edupage 12/10/98. (edupage@franklin.oit.unc.edu).

Janis, I. L. (1972). *Victims of groupthink: A psychological study of foreign policy decisions and fiascos.* Boston: Houghton-Mifflin.

Johnson, R. (1997). *User-centered technology: A rhetorical theory for computers and other mundane artifacts.* Albany: State University of New York Press.

Karjala, D. (1999). *What Are the Issues in Copyright Term Extension—and What Happened?* www.public.asu.edu/~dkarjala/what.html

Kohl, J. R. (1999). Improving translatability and readability with syntactic cues. *Technical communication 46*(2), 149–166.

Kostelnick, C., & Roberts, D. D. (1998). *Designing visible language: Strategies for professional communicators.* Boston: Allyn & Bacon.

Lavin, M. R. (1992). *Business information: How to find it, how to use it* (2nd ed.). Phoenix: Oryx Press.

Lynch, P. J., & Horton, S. (1999). *Web style guide: Basic design principles for creating Web sites.* New Haven: Yale University Press.

Miles, T. H. (1989). The memo and "disinformation": Beyond format and style. *Issues in Writing, 2*(1), 42–60.

Munger, D. D., Anderson, D., Bret, D., Busiel, C., & Paredes-Holt, B. (2000). *Researching online* (3rd ed.). New York: Longman.

Pace, R. C. (1988). Technical communication, group differentiation, and the decision to launch the space shuttle *Challenger. Journal of Technical Writing and Communication, 18*(3), 207–220.

Patry, W. F. (1985). *The fair use privilege in copyright law.* Washington, DC: Bureau of National Affairs.

Perelman, L. J. (1993, October 25). How hypermation leaps the learning curve. *Forbes* ASAP, 78+.

Petroski, H. (1996). *Invention by design.* Cambridge, MA: Harvard University Press.

Presidential commission. (1986). *Report to the president on the space shuttle Challenger accident.* Vol. 1. Washington: Government Printing Office.

Rosenfeld, L., & Morville, P. (1998). *Information architecture for the world wide web.* Cambridge, MA: O'Reilly.

Rubin, J. (1994). *Handbook of usability testing: How to plan, design, and conduct effective tests.* New York: Wiley.

Ruggiero, V. R. (1998). *The art of thinking* (5th ed.). New York: Addison.

Schenk, M. T., & Webster, J. K. (1984). *Engineering information resources.* New York: Decker.

Seglin, J. L. (1998, July). Would you lie to save your company? *Inc.,* 53–57.

Steinberg, S. (1994, July). Travels on the net. *Technology Review,* 20–31.

Steinberg, S. H. (1996). *Five hundred years of printing* (Rev. ed.). London: British Library and Oak Knoll Press.

Strong, W. S. (1993). *The copyright book: a practical guide.* Cambridge, MA: MIT Press.

Tufte, E. R. (1990). *Envisioning information.* Cheshire, CT: Graphics Press.

———. (1992). *The visual display of quantitative information.* Cheshire, CT: Graphics Press.

Walker, J. R., & Ruszkiewicz, J. (Eds.). (2000). *Writing*@online.edu. New York: Longman.

Weiner, D. B. (1999, July). Genetic vaccines. *Scientific American,* 50–57. (Human trial of DNA vaccines table.)

Wickens, C. D. (1992). *Engineering Psychology and Human Performance* (2nd ed.). New York: Harper.

Wilford, J. N. (1999, April 6). When no one read, who started to write? *New York Times,* pp. D1, D2.

Winsor, D. (1988). Communication failures contributing to the *Challenger* accident: An example for technical communication. *IEEE Transactions on Professional Communication, 31*(3), 101–107.

ZDNET. (2000). *Intel pill: is Big Brother inside?* www.zdnet.com/zdhelp/stories/main/0,5594,2214831,00.html

Credits

Chapter 2

Figure 2.1, Reprinted with permission from Medtronic, Inc. © Medtronic, Inc. Figure 2.2, Black & Decker Instruction Manual © 1993 The Black & Decker Corporation. Figure 2.3, Reproduced with permission of Schering Corporation. All rights reserved. Figure 2.4, *Physicians Desk Reference* copyright 2000 published by Medical Economics Company, Inc., Thomson Healthcare.

Chapter 3

Figure 3.3, Courtesy Casady & Greene. Figures 3.4 and 3.5, Copyright IBM Corporation, permission to reproduce granted by IBM.

Chapter 5

Figure 5.1, Courtesy of PBS, www.pbs.org. Figure 5.2, Used with permission. Figure 5.3a, Pesticide Management Education Program at Cornell University, Ithaca, NY.; 5.3b, Courtesy of B. L. Travel, Poultney, Vermont. Figure 5.4, Screen shot reprinted by permission from Microsoft Corporation. Figure 5.5, Screen shot reprinted by permission from Microsoft Corporation.

Chapter 6

Figure 6.3, Copyright © 1993, 2000 ACM. Used by permission. Figure 6.4, Reprinted with permission from the Society for Technical Communications, Arlington, VA, U.S.A.

Chapter 7

Figure 7.1, Reprinted with permission from *Technical Communication,* the journal of the Society for Technical Communication. Figure 7.3, Reprinted with the permission of the Regents of the University of Minnesota. © 2000 Regents of the University of Minnesota. Figure 7.4, Amazon.com is the registered trademark of Amazon.com, Inc.

Chapter 8

Figure 8.9, Reprinted with permission from Symantec Corporation. Figure 8.10, Manual page reprinted by permission from Microsoft Corporation. Figure 8.12, Reprinted with permission: SAS Institute Inc., *Corporate Identity: Written Style,* Second Edition, Cary, NC: SAS Institute Inc., 1989, 212 pp.

Chapter 9

Figure 9.2, The Center for Analysis and Prediction of Storms at the University of Oklahoma. Figure 9.3, National Oceanic and Atmospheric Administration, National Climatic Data Center, Asheville, NC. Figure 9.6, Courtesy of Werner Paddles. Figure 9.10, United States Historical Census Data Browser: Author/Host: University of Virginia Library. Geospatial and Statistical Data Center, URL: fisher.lib.virginia.edu/census. Figure 9.13, Courtesy of Intuit. Figure 9.14, Screen shot reprinted by permission from Microsoft Corporation. Figure 9.15, Courtesy of AEC Software, Inc. © 1996. Figure 9.19, Reprinted by permission from Slim Films. Figure 9.20, Reprinted by permission of Daniels & Daniels. Figure 9.21, Reprinted by permission from Slim Films. Figure 9.23, Courtesy of Krups, Inc. Figure 9.24, Capital Features/The Image Works. Figure 9.26, Courtesy of BestBuy.com

Chapter 10

Figure 10.1, Netscape Messenger screen shot © 2000 Netscape Communications Corporation. Used with permission. Figure 10.4, Screen shot reprinted by permission from Microsoft Corporation. Figure 10.9, Screen shot reprinted by permission from Microsoft Corporation.

Chapter 11

Figure 11.1, Portions reprinted with permission from IEEE Std. C62.1-1989, IEEE Standard for Gapped Silicon-Carbide Surge Arresters for AC Power Systems. Copyright 1989 by IEEE. The IEEE disclaim any responsibility or liability resulting from the placement and use in the described manner. Figure 11.2, Copyright © 1998–99 Hewlett-Packard Company. Reproduced with permission. Figure 11.4, Reprinted with permission of Iomega Corporation. Figure 11.5, Copyright © 1999 Hewlett-Packard Company. Reproduced with permission. Figure 11.6, Reprinted with the permission of the Regents of the University of Minnesota. © 2000 Regents of the University of Minnesota. Figure 11.7, Courtesy of Pittsburgh Corning Corporation. Figure 11.8, Courtesy of Casady & Greene. Figure 11.9, Reprinted with permission from Symantec Corporation. Figure 11.10, Reprinted with permission from Symantec Corporation. Figure 11.11, Reprinted with permission from Medtronic, Inc. © Medtronic, Inc. Figure 11.12, Courtesy of Toyota Motor Sales, U.S.A., Inc.

Chapter 12

Figure 12.4, Reprinted by permission from George Retseck.

Index

Page numbers followed by italicized letter *f* indicate figures.

A

Abbreviations, 299
Abstracts, 65, 269
Accessibility of information, 4–5, 6,
    148–149
Access points, in manuals, 233, 234*f*
Accuracy of communication, 5
Action plan, 207–208
Action verbs, 226, 272
Active voice, 38–39, 222, 231
Addressing letters, 198
Adobe Framemaker, 139
Adobe Illustrator, 169
Adobe InDesign, 170
Adobe Photoshop, 111, 169
Affect vs. effect, 293
All capital letters
    in email, 186
    use of, 126–127
Almanacs, 63
Among vs. between, 293
Analytical reports, 7, 260*f*–268*f*
APA documentation style, 316–324
Apostrophe, 287
Appendices, 270
Area graph, 154, 155*f*
Ascender of letter form, 127
As if vs. like, 294
Assembly instructions, 224
Asynchronous communication, 79
Attachments of email, 79, 79*f*, 186
Attention line, 197
Audience
    color and, 170–171
    feedback from, 33–35, 37
    general *vs.* specific, 18, 20*f*–21*f*
    importance of, in technical writing, 15
    length of documents and, 24

point of view of, 41–42, 235–236
primary, 19
relevance of information and, 5
secondary, 19
tools for understanding, 24
typical, 26–27
viewing page, 128–129
visuals and, 149–150, 150*f*, 151*f*
of Web pages, 72–73
word choice and, 24
Audience analysis, 18–19, 23
    brief instructions, 222
    definitions and descriptions, 247
    email, 184
    letters, 190–191
    long reports, 259
    manuals, 232–233
    memos, 187
    oral presentations, 206–207
    procedures, 227
    proposals, 272
    short reports, 200
    specifications, 216–218
    technical marketing material, 238
Audience/purpose interview, 23–24

B

Background colors, 170
Band graph, 154, 155*f*
Bar graphs, 154
    deviation, 157, 158*f*
    multiple, 157, 157*f*
    simple, 154–156, 156*f*
Between vs. among, 293
Biased language, 41
Bibliographic databases, 61
Bibliographic Retrieval Services (BRS),
    61

Bibliographies, 62
Bibliography programs, 60
Body section of documents
    emails, 186
    fonts for, 122
    letters, 195, 196*f*
    long reports, 270
    memo, 189
    oral presentations, 208–209
    short reports, 201
Bold letters, use of, 126
Book indexes, 64
Bookmarking files, 60
Books
    APA entries for, 318–320
    MLA entries for, 306–309
Boolean logic, in keyword search, 57–58
Brief definitions/descriptions, 255
Brief instructions, 222–226
    audience and purpose analysis, 222
    components of, 225
    types of, 222–224
    usability considerations, 225–226
Brief reports, 188–189
Brochures
    audience of, 15, 16*f*
    as technical marketing materials,
        238, 241*f*, 242*f*
Browsers, 57, 172
BRS. *See* Bibliographic Retrieval Services
Bulleted lists, 130–131, 132*f*, 288–290
    parallel structure of, 296–297
    in short reports, 201

C
Calls for proposals (CFPs), 8, 273
Capitalization, 299–300
Card catalog, 62
Causal reports, 259
Cautions, 230
CD-ROMs
    copyright and, 113
    designing, 143
    as research tools, 60–61
CFPs. *See* Calls for proposals
Challenger, 88–89
Chapters, in manuals, 233–235
"Chartjunk," 172

Charts, 157–161. *See also specific chart*
Chronological sequence, 258
Chunking information, 43, 45, 72, 75, 131
Citation indexes, 64
Clarity of communication, 5, 6, 199, 206
Clip art, copyright-free, 112
Closed-ended questions, 53
Closing of letters, 195
Codes of ethics, 99–101, 99*f*, 100*f*
Cold calls, 238, 273
Collective nouns, 295
Colon, 288
Color, using, 170–171, 171*f*
    cost and, 170, 171
    cultural issues, 171
    in technical marketing materials,
        243
Comma, 37–38, 286–287
Comma splice, 294
Communication situations
    complex, 182, 246–284
    everyday, 182, 183–211
    product-oriented, 182, 215–243
Compact discs. *See* CD-ROMs
Comparative reports, 258
Completeness of communication, 5
Complex communication situations,
    182, 246–284
Computer projection software, 209
Computer-supported cooperative work
    systems, 81
Conclusion section of documents
    long reports, 270
    oral presentations, 209
Concreteness of communication, 5
Consequences, 92
Context of communication, analyzing,
    22, 23
Context-sensitive online documentation,
    77, 232
Continual vs. continuous, 293
Cookies, 114–115, 114*f*, 116
Copyright, 105–113
    electronic technologies and, 111–113
    establishing, 107
    ethical issues, 92–93
    infringing, 107–110, 111, 172
    origin of, 106

Copyright-free clip art, 112
Copyright holders, rights of, 107, 108*f*
CorelDraw, 169
Corel WordPerfect, 139, 170
Corporate authors, 109, 110
Corporate style guides, 24, 128,
      137–139, 138*f*, 140*f*–141*f*
"Courtesy copy," 197
Cover letter, 192, 193*f*
Cultural differences
    color and, 171, 173
    exploiting, 96
    privacy laws, 115–116
    translating and, 42
Cutaway diagrams, 163, 165*f*

D
Dangling modifier, 295–296
Databases
    bibliographic, 61
    factual, 61
    full-text, 61
Date, in letters, 194
Definitions and descriptions, 247–258
    audience and purpose analysis, 247
    components of, 255–256
    types of, 255
    usability considerations, 256–258
Descender of letter form, 127
Descriptions. *See* Definitions and
    descriptions
Designing documents. *See* Page and doc-
    ument design
Deviation bar graphs, 157, 158*f*
Diagrams, 162–163, 163*f*, 225, 236. *See*
    also specific diagram
DIALOG, 61
Dictionaries, 63
Direct approach, 198, 198*f*
Directories, 63
Disinterested vs. uninterested, 293
Display booths, 238
Display typefaces, 124, 124*f*
Distance education, 81
Distribution notations, 197
Document design. *See* Page and docu-
    ment design

Document genre, purposes of communi-
    cation and, 24
Downloading files, 60

E
Effect vs. affect, 293
Electronic age, 9–10
Electronic magazines, 56
Electronic mail. *See* Email
Electronic messages, copyright and, 113
Electronic note cards, 60
Electronic sources
    APA entries for, 323–324
    MLA entries for, 313–316
Email, 8, 79–81
    audience and purpose analysis, 184
    components of, 185–186
    copyright and, 113
    inquiries via, 56
    types of, 184–185, 185*f*
    usability considerations, 186–187
Embedded lists, 288
Emoticons, 187
Enclosure notations, 197
Encyclopedias, 62
Endnotes, 305
Engineers, 26
Ethical considerations, 11, 86–101. *See*
    also Unethical communication
    case examples of, 87–90
    in graphics, 96–97
    in instructions, 97–98
    in memos, 97
    in oral presentations, 98
    in proposals, 98
    in reports, 98
    responding to, 99–101
    types of choices, 90–92
    in Web pages, 97
Ethical relativism, 91–92
Ethical standards *vs.* legal standards,
    92–94
Etymology, 255
Everyday communication situations,
    182, 183–211
Executive summary, 269
Executives, 26

Expanded definitions/descriptions, 255
Experiments, 54
Exploded diagrams, 163, 164*f*

F
Factual databases, 61
Fair use doctrine, 92, 109–110, 109*f*, 111
Farther vs. further, 293
Faulty modification, 295–298
Feasibility reports, 259
Fewer vs. less, 293
Files, downloading, 60
Flaming, 56, 80, 186
Flowcharts, 159
Focus groups, 23, 24
FOIA. *See* Freedom of Information Act
Fonts, 122–124
    display typefaces, 124, 124*f*
    message sent by, 125
    mixing and matching, 125–127, 126*f*
    readability and, 122, 127
    sans serif type, 74, 75*f*, 123–124, 123*f*
    serif type, 74, 122, 123*f*
    size of, 127
    standard formats, 127–128
Footnotes, 305
Forecasting statements, 298
Freedom of Information Act (FOIA), 53,
    65
Frequently Asked Questions (FAQs), 240
Full-text databases, 61
Functional sequence, 258
Functional specs, 218, 219*f*
Further vs. farther, 293

G
Gantt chart, 159, 160*f*
Gender, Internet use and, 80–81
Global issues. *See* Cultural differences
Glossary, 270
Government publications, access tools
    for, 65
Government standards, 218
Grammar, 37–42, 80, 288–301
Grammar checkers, 42, 190
Grammatically parallel headings, 134
Grant proposal, 273, 275*f*–276*f*
Graphics. *See* Visuals

Graphics software, 169
Graphs, 154–157, 172. *See also specific
    graph*
Grid structure, 129, 130*f*
Groupthink, 95
Guides to literature, 63
Gutenberg diagram, 129, 129*f*

H
Handbooks, 63
Handouts, 209, 210
Hard copy research, 62–65
Header, of email, 185–186
Headings, 45, 131–134
    appropriate size, 134
    of CD-ROMs, 143
    consistency of, 134
    in form of questions, 44–45, 134
    grammatically parallel, 134
    informative, 271–272
    left-margin cueing area, 134, 135*f*
    levels, 132–133
    in long reports, 271–272
    in online help, 143
    running heads, 134, 136*f*, 233
    in short reports, 201, 206
    on Web pages, 142–143
Hierarchy, email and, 80
Horton, William, 148
Hypertext format, 72
Hypertext markup language (HTML),
    73, 142
Hyphenation, 301

I
Icons, 165–166
Ideals, 91
Illustrations, 161, 162*f*, 236
Images, Web-based, 169–170
Imaging, medical, 169
Imperative voice, 226
Imply vs. infer, 293
Indefinite pronouns, 295
Indexes
    book, 64
    citation, 64
    creating, 137

in manuals, 233
online, 77
patent, 64
periodical, 64
technical report, 64
Indirect approach, 198, 198*f*
Industry standards, 218
Infer vs. imply, 293
Infinitive, introductory sentence of lists
    ending with, 289–290
Information
    accessible, 4–5, 6, 148–149
    chunking, 43, 45, 72, 75, 131
    relevant, 5–7, 149
    usable. *See* Usability of information
Information overload, 10
Information plan, 33, 34*f*, 35
Informative interviews, 51–52
Informative presentation, 207
Inquiry letter, 192
Instructions, 7
    assembly, 224
    brief. *See* Brief instructions
    ethical considerations in, 97–98
    long, 227–228, 229*f*
    safety, 15, 17*f*
    wordless, 166, 167*f*, 224, 224*f*
Intellectual property, 106
Interface design, 77–78
International issues. *See* Cultural differ-
    ences
International Standards Organization
    (ISO), 165, 172
Internet
    copyright and, 111–112
    privacy and, 114–116
Internet research, 54–60
Internet specs, 218
Interview(s)
    audience/purpose, 23–24
    informative, 51–52
Introduction section of documents
    long reports, 270
    oral presentations, 208
    proposals, 277
    specifications, 218–221
Introductory sentence of lists,
    289–290

ISO. *See* International Standards
    Organization
Italics, use of, 125, 287

J
Jargon, useless, 41
Job application, cover letter for, 192, 193*f*

K
Kant, Immanuel, 90
Keyword search, 57–58

L
Language
    audience's level of experience and, 198
    biased, 41
    clear and concise, 257–258
    convincing, 280
    dynamic, 243
Large color documents, 239–240
Lawyers, 27
Layered communication, 35, 36*f*, 227, 232
Lay vs. lie, 293
Left-margin cueing area, 134, 135*f*
Legal standards *vs.* ethical standards, 92–94
Length of documents, audience and, 25
Less vs. fewer, 293
Letter(s), 190–199
    audience and purpose analysis, 190–191
    components of, 192–197
    as technical marketing materials, 239
    types of, 191–192
    usability considerations, 198–199
Letterheads, 122, 194
Lie vs. lay, 293
Like vs. as if, 294
Line graphs
    multiline, 154, 155*f*
    simple, 154, 154*f*
    unethical use of, 97
    without labels, 148*f*
Line length, of Web pages, 74–75, 76*f*
Links, organizing, 73
Lists, 288–290
    bulleted. *See* Bulleted lists
    embedded, 288
    numbered. *See* Numbered lists
    parallel structure of, 296–297

Lists (*continued*)
    vertical. *See* Bulleted lists; Numbered
        lists
Listservs, 55–56
Long instructions, 227–228, 229*f*
Long reports, 258–272
    audience and purpose analysis, 259
    components of, 269–270
    types of, 259
    usability considerations, 271–272

M
Macromedia Director, 169
Mailing list, computer-operated, 55–56
Managers, 26–27
Manuals, 7, 232–236
    audience and purpose analysis,
        232–233
    components of, 233–235
    types of, 233
    usability considerations, 235–236
Maps, 166, 168*f*
Margins, using, for commentary, 45
Marketing surveys, 24
Markup languages, 73, 139–142
Medical imaging, 169
Meeting minutes, 188, 200–201
Memos, 8, 187–190, 188*f*
    audience and purpose analysis, 187
    components of, 189
    ethical considerations in, 97
    short reports beginning with, 201
    types of, 187–189
    usability considerations, 189–190
Message body, of email, 186
Microforms, 65
Microsoft Excel, 170
Microsoft Internet Explorer, 172
Microsoft PowerPoint, 169, 206
Microsoft Project, 170
Microsoft Word, 139, 170, 192
Mill, John Stuart, 91
Misplaced modifier, 296
MLA documentation style, 305–316
Moderated newsgroups, 55
Modifiers
    dangling, 295–296
    misplaced, 296

Multiline graph, 154, 155*f*
Multimedia, copyright and, 113
Multiple-bar graph, 157, 157*f*

N
Netiquette, 56, 80
Netscape Navigator, 172
Newsfeed newsgroups, 55
Newsgroups
    moderated, 55
    newsfeed, 55
    unmoderated, 55
Newspapers, fonts used in, 122, 128
Nominalizations, 39
Nonlinear readers, 72
Nouns
    collective, 295
    introductory sentence of lists ending
        with, 289
    nominalizations, 39
    symbols representing, 165–166, 166*f*
    too many in a row, 39–40
Numbered lists, 130–131, 132*f*, 291
    instructions written in, 225
    parallel structure of, 297
    in procedures, 230
    in short reports, 201
Numbers, use of, 300–301

O
Obligations, 91
Online Computer Library Center
    (OCLC), 61
Online documentation, 77, 78*f*
Online help, 232
    definitions in, 255
    designing, 143
Online retrieval services, 61–62
Open-ended questions, 53
Oral presentations, 206–211
    audience and purpose analysis,
        206–207
    components of, 208–210
    ethical considerations in, 98
    types of, 207–208
    usability considerations, 210–211
Organizational charts, 159, 161*f*
Organizational publications, 53–54

Organizational style guides, 24, 128, 137–139, 138f, 140f–141f
Organizing documents, 5
 usability and, 37–47
 Web pages, 73, 74f
Outline, 132–133, 211, 211f, 269, 269f
Overediting, 40
Overhead transparencies, 209
Overview
 in brief instructions, 225
 creating, 43–44, 43f
 in manuals, 233
 in procedures, 230

P
Page(s), visual hierarchy of, 128–129
Page and document design, 120–143
 of short reports, 201
 typography. *See* Typography
 unethical use of, 97
 usability and, 33–35, 42
 of Web pages. *See* Web page design
Page layout software, 139–142
PageMaker, 139
Paragraphs, transitions within and between, 297–298
Parallel structure, 40, 296–297
Parenthetical references, 305–306, 316–317
Passive voice, 38–39
Patent indexes, 64
Patent law, 106
Pentium III Chip, 89–90
Percent vs. percentage, 294
Period, 286
Periodical indexes, 64
Periodicals
 APA entries for, 320–321
 MLA entries for, 309–310
Personal data, collecting, in U.S. *vs.* in Europe, 116
Personal observation, 54
Persuasive presentation, 207
Photocopiers, copyright and, 111
Photographs, 166, 168f
Photography software, 169
Picas, 127
Pictogram, 161, 162f, 172

Pictures. *See* Visuals
Pie chart, 158, 159f, 160f
Plagiarism, 93, 303
Planning documents, usability and, 31–33
Planning proposal, 273, 274f
Points, 127
Preposition, introductory sentence of lists ending with, 289
Presentations, ethical considerations in, 98
Presentation software, 81, 169
Primary audience, 19
Primary research, 51–54
Principle vs. principal, 294
Printing files, 60
Privacy, 93–94, 113–117. *See also* Copyright
 in cyberspace, 114–116
 documentation and, 116–117
 Pentium III chip, 89–90
 videotapes and, 117
Privacy laws, compliance with, 115–116
Privacy statements, 115, 115f
Procedures, 7, 226–231
 audience and purpose analysis, 227
 components of, 228–231
 types of, 227–228
 usability considerations, 231
Product-oriented communication situations, 182, 215–243
Progress reports, 200, 204f–205f
Project management software, 170
Pronoun-antecedent agreement, 295
Pronouns, indefinite, 295
Proposals, 8, 272–280
 audience and purpose analysis, 272
 components of, 277–278
 ethical considerations in, 98
 types of, 273–277
 usability considerations, 278–280
 vs. technical marketing material, 236
Public, as audience, 27
Public domain, 109
Public records, 53–54
Punctuation, 37–38, 286–288
 of lists, 131

Purpose analysis, 19, 22, 23
  brief instructions, 222
  definitions and descriptions, 247
  email, 184
  letters, 190–191
  long reports, 259
  manuals, 232–233
  memos, 187
  oral presentations, 206–207
  procedures, 227
  proposals, 272
  short reports, 200
  specifications, 216–218
  technical marketing material, 238
Purpose interview. *See* Audience/purpose interview
Purposes of communication, 15
  document genre and, 24
  relevance of information and, 5
  typical, 26–27
  Web pages, 19, 22

Q
Quark, 139
Questionnaires, 52–53
Questions
  headings in form of, 45, 134
  open-ended *vs.* closed-ended, 53
Quick reference cards, 7, 222–224, 223*f*
Quotation marks, 287, 303
Quoting. *See* Copyright; Sources

R
Readability, fonts and, 122, 127
Readers. *See* Audience
Reasonable criteria for ethical judgment, 91–92
Recommendation reports, 7, 200, 202*f*–203*f*
Reference lists, 110. *See also* Works cited entries
  APA style, 317–324
  in manuals, 235
Relativism, ethical, 91–92
Relevance of information, 5–7, 149
Reports, 7–8
  brief, 188–189
  ethical considerations in, 98

long. *See* Long reports
  short. *See* Short reports
Requests for proposals (RFPs), 8, 273
Research, 50–65
  card catalog, 62
  CD-ROMs, 60–61
  developing information plan and, 33
  hard copy, 62–65
  Internet, 54–60
  online retrieval services, 61–62
  primary, 51–54
  for short reports, 201
Research Libraries Information Network (RLIN), 61
Research proposal, 273, 275*f*–276*f*
Revising documents, 35, 37
RFPs. *See* Requests for proposals
RLIN. *See* Research Libraries Information Network
Running heads, 134, 136*f*, 233
Run-on sentences, avoiding, 292

S
Safety instructions, 15, 17*f*
Sales presentation, 208
Sales proposal, 273–277, 278*f*
Salutation, in letters, 195
Sample of surveys, 52
Sans serif typefaces, 74, 75*f*, 123–124, 123*f*
Scanners, copyright and, 111
Scientists, 26
Screen shots, 236, 237*f*
Search engines, 57–58
Secondary audience, 19
Semicolon, 286, 292
Sentence fragments, avoiding, 290–291
Sequence
  chronological, 258
  functional, 258
  spatial, 258
Series comma, 37–38
Serif typefaces, 74, 122, 123*f*
Sexist language, in salutation, 195
SGML. *See* Standardized general markup language
Short reports, 199–206
  audience and purpose analysis, 200

components of, 201
types of, 200–201
usability considerations of, 201, 206
Signature, 195, 197
Simple bar graph, 154–156, 156*f*
Simple line graphs, 154, 154*f*
   unethical use of, 97
   without labels, 148*f*
Situations. *See* Communication situations
Slides, 209, 209*f*, 211
Smiley faces, 187
Software(s)
   computer projection, 209
   creating visuals with, 169–170
   page layout, 139–142
   presentation, 81
   word-processing, 139–142
Software charts, 159
Solicited proposal, 273
SOPs. *See* Standard operating procedures
Sources
   documenting, 110, 302–324. *See also*
     Copyright
   evaluating, 58–59
Spamming, 56
Spatial sequence, 258
Specifications, 216–222
   audience and purpose analysis,
     216–218
   components of, 218–221
   types of, 218
   usability considerations, 221–222
Speed, 72, 80
Spelling, 80, 81, 303
   email, 186
   letters, 199
   slides, 211
Spelling checkers, 42, 190
Spreadsheet software, 170
Standardized general markup language
   (SGML), 142
Standard operating procedures (SOPs),
   7, 226, 227, 228*f*
Style, 37–42
Style guides, corporate, 24, 128,
   137–139, 138*f*, 140*f*–141*f*
Subchunks, 132
Subheadings, 132–133

Subject directories, 57
Subject line, 197
Subject-verb agreement, 294–295
Sumerians, writing of, 4
Surveys, 52–53
Symbols, 165–166, 166*f*, 230

T
Table of contents
   creating, 137
   in manuals, 233, 234*f*
Tables, 150–151, 152*f*, 153*f*
Tags. *See* Markup languages
Target population, 52
Task analysis, 31–33
Task-oriented technical communication,
   31
Teamwork *vs.* groupthink, 95
Technical brochure, 15, 16*f*
Technical communication
   audience of. *See* Audience
   characteristics of, 4–7
   context of, 22, 23
   in electronic age, 9–10
   ethical dimensions of. *See* Ethical
     considerations
   history of, 4
   layered, 35, 36*f*, 227, 232
   purposes of. *See* Purposes of commu-
     nication
   situations. *See* Communication situa-
     tions
   societal dimensions of, 10
   task-oriented, 31
   types of, 7–9
   user-centered, 15
Technical communicators, 9
Technical marketing material, 236–243
   audience and purpose analysis, 238
   components of, 240
   types of, 238–240
   usability considerations, 240–243
   vs. proposals, 236
Technical report indexes, 64
Telecommuting, 81
"Term-class-features" method, 255
Terminology, consistent, 6

Testing, of early versions of documents, 33–35
Three Mile Island, 87–88
Time constraints, with presentations, 210
Title of documents
    brief instructions, 225
    procedures, 228
Trademark law, 106
Training session, 207
Translation, writing for, 42
Transmittal letter, 191
Transmittal memo, 187–188
Tree charts, 159–161, 161*f*
Truncation, 58
Typist's initials, 197
Typography, 121–128. *See also* Fonts
    of CD-ROMs, 143
    of online help, 143
    unethical use of, 97
    of Web pages, 74, 75*f*, 124, 142

U
Underlining, use of, 287
Unethical communication
    examples of, 94–96
    legal, 92–94
Uninterested vs. disinterested, 293
Unmoderated newsgroups, 55
Unsolicited proposal, 273
URLs, copying, 60
Usability of information, 5, 30–45
    after release, 35–37
    in brief instructions, 225–226
    in definitions and descriptions, 256–258
    in email, 186–187
    in letters, 198–199
    in long reports, 271–272
    in manuals, 235–236
    in memos, 189–190
    in oral presentations, 210–211
    during planning stages, 31–33
    in procedures, 231
    in proposals, 279–280
    in short reports, 201, 206
    in specifications, 221–222
    in technical marketing material, 240–243

visuals and, 149
during writing and design process, 33–35
writing and organizing information for, 37–45
Usenet, 55
User-centered technical communication, 15
User preference documents, 24
Users. *See* Audience
Utilitarianism, 91

V
Verbs
    action, 226, 272
    agreeing with subject, 294–295
    introductory sentence of lists ending with, 289
    symbols representing, 165–166, 166*f*
Vertical lists. *See* Bulleted lists;
        Numbered lists
Videotapes, privacy and, 117
Visual hierarchy of pages, 128–129
Visualization, 169
Visual noise, 170, 172
Visuals, 147–173
    charts, 157–161
    color, 170–171, 171*f*, 173
    cultural considerations, 172–173
    in definitions and descriptions, 257
    diagrams, 162–163, 163*f*
    for different audiences, 149–150, 150*f*, 151*f*
    ethical considerations in, 96–97
    ethics and, 172
    graphs, 154–157
    icons, 165–166
    illustrations, 161, 162*f*
    in long reports, 271
    in manuals, 236
    maps, 166, 168*f*
    in oral presentations, 209, 209*f*
    photographs, 166, 168*f*
    in short report, 201
    symbols, 165–166, 166*f*
    tables, 150–151, 152*f*, 153*f*
    in technical marketing materials, 240, 243

use of, 149
wordless instruction, 166, 167f

W

Warnings, 230
Web-based images, 169–170
Web page(s)
   copyright and, 111–112
   definitions on, 255
   ethical considerations, 97
   hypertext format of, 72
   privacy and, 114–116
   purposes of, 19, 22
   research on, 57–60
   as technical marketing materials,
      238, 239f
Web page design, 71, 142–143
   design issues, 73–75
   headings, 142–143
   technical issues, 75–77

typography, 74, 75f, 124, 142
   writing issues in, 72–73
White space, proper use of, 42, 130
Word choice, audience and, 25
Wordless instructions, 166, 167f, 224, 224f
Word-processing software
   letter templates, 192, 194f
   page layout, 139–142
   working with visuals, 170
Wordy phrases, 40
Workplace research, 54
Works cited entries, 306–316. *See also*
      Reference lists
Works-for-hire doctrine, 110
Writing documents
   usability and, 33–35, 37–45
   Web pages, 72–73

Z

Zines, 5